Smart Antennas, Electromagnetic Interference and Microwave Antennas for Wireless Communications

RIVER PUBLISHERS SERIES IN COMMUNICATIONS AND NETWORKING

The "River Publishers Series in Communications and Networking" is a series of comprehensive academic and professional books which focus on communication and network systems. Topics range from the theory and use of systems involving all terminals, computers, and information processors to wired and wireless networks and network layouts, protocols, architectures, and implementations. Also covered are developments stemming from new market demands in systems, products, and technologies such as personal communications services, multimedia systems, enterprise networks, and optical communications.

The series includes research monographs, edited volumes, handbooks and textbooks, providing professionals, researchers, educators, and advanced students in the field with an invaluable insight into the latest research and developments.

Topics included in this series include:-

- Multimedia systems;
- Network architecture;
- Optical communications;
- Personal communication services;
- Telecoms networks;
- Wifi network protocols.

For a list of other books in this series, visit www.riverpublishers.com

Smart Antennas, Electromagnetic Interference and Microwave Antennas for Wireless Communications

Editors

S. Kannadhasan
Cheran College of Engineering, India

R. Nagarajan
Gnanamani College of Technology, India

Alagar Karthick
KPR Institute of Engineering and Technology, India

Aritra Ghosh
University of Exeter, UK

Published 2022 by River Publishers
River Publishers
Alsbjergvej 10, 9260 Gistrup, Denmark
www.riverpublishers.com

Distributed exclusively by Routledge
4 Park Square, Milton Park, Abingdon, Oxon OX14 4RN
605 Third Avenue, New York, NY 10017, USA

Smart Antennas, Electromagnetic Interference and Microwave Antennas for Wireless Communications / S. Kannadhasan, R. Nagarajan, Alagar Karthick and Aritra Ghosh.

Routledge is an imprint of the Taylor & Francis Group, an informa business

ISBN 978-87-7022-776-6 (print)
ISBN 978-10-0084-642-3 (online)
ISBN 978-1-003-37323-0 (ebook master)

Contents

Preface xv

List of Contributors xvii

List of Figures xix

List of Tables xxv

List of Abbreviations xxvii

Chapter 1
Speech Signal Extraction from Transmitted Signal Using Multilevel Mixed Signal 1
R. Kabilan, and R. Ravi

1.1 Introduction 1
1.2 Literature Survey 2
1.2.1 The fast ICA algorithm revisited: Convergence analysis 2
1.2.2 FPGA implementation of IC algorithm for blind signal separation and noise cancelling 3
1.2.3 Subjective comparison and evaluation of speech enhancement algorithms. 3
1.3 Proposed Systems 3
1.3.1 FASTICA using symmetric orthogonalization 3
1.3.2 FPGA implementation 4
1.4 Results and Discussion 5
Output Waveform 5
1.5 Conclusion. 9
References 10

Chapter 2
High Performance Fiber-Wireless Uplink for CDMA 5G Networks Communication 13
R. Ravi, R. Kabilan, and S. Shargunam
2.1 Introduction 14
2.2 Proposed Method 15
2.2.1 OFDM 17
2.2.2 OFDMA 18
2.2.3 CDMA 18
2.2.4 Optical fiber channel 18
2.2.5 The disadvantages of the existing system 18
2.3 Results and Discussion 18
2.3.1 Inference 1 18
2.3.2 Inference 2 19
2.3.3 Inference 3 19
2.3.4 Inference 4 20
2.3.5 Inference 5 20
2.3.6 Inference 6 22
2.3.7 Inference 7 23
2.4 Conclusion 23
References 25

Chapter 3
Improving the Performance of Cooperative Transmission Protocol Using Bidirectional Relays and Multi User Detection 29
R. Kabilan, R. Ravi, and S. Shargunam
3.1 Introduction 30
3.2 Components of Communication System 30
3.3 Proposed System 31
3.4 System Design and Development 32
3.4.1 Input design 32
3.4.2 Feasibility analysis 32
3.4.2.1 Operational feasibility 32
3.4.2.2 Technical feasibility 32
3.4.2.3 Economical feasibility 32
3.4.2.4 Project modules 32
3.5 Output Design 33
3.5.1 Animator output 34
3.5.2 Initialization of nodes 34

3.5.3 Node 1 starts transmitting data 34
3.5.4 Finding shortest path 35
3.5.5 Transmission of data through relay node 36
3.5.6 Node 8 starts transmitting data 37
3.5.7 Loss of packets 37
3.5.8 Transmissions of data from node 7 to node 6 37
3.5.9 Transmision of data from node 2 to node 4 37
3.5.10 Transmission of data bidirectionally 38
3.5.11 Completion of transmission from node 8 to 0. 39
3.5.12 Coverage provided by dynamic base station 40
3.5.13 Retransmission of dropped packets 41
3.5.14 Reception of acknowledgement 41
3.5.15 X graph for lifetime 41
3.5.16 X graph for output. 41
3.6 Conclusion. 43
References 44

Chapter 4
Joint Relay-source Escalation for SINR Maximization in Multi Relay Networks and Multi Antenna **47**
R. Ravi, R. Kabilan, S. Shargunam, and R. Mallika Pandeeswari
4.1 Main Text 48
4.2 Proposed System 49
4.2.1 System model 49
4.2.2 SINR maximization under relay transmit power and source constraints 50
4.2.3 Source-relay transmit power minimization under QoS constraints 50
4.2.4 Computation of relay precoder 51
4.2.5 Feasibility of the problem 51
4.3 Advantage 51
4.4 Application 51
4.5 Result and Discussion 51
4.5.1 Tools used 51
4.5.2 Simulated results 52
4.6 Conclusion 56
References 57

Chapter 5
VLSI Implementation on MIMO Structure Using Modified Sphere Decoding Algorithms 59
R. Kabilan, R. Ravi, S. Shargunam, and R. Mallika Pandeeswari
5.1 Introduction 60
5.2 Proposed Methodology 60
5.2.1 VB decoding algorithm 61
5.2.2 SE decoding algorithm 62
5.2.3 SOC architecture on FPGA 64
5.3 Result and Conclusion. 64
5.4 Conclusion. 68
References 68

Chapter 6
Overcrowding Cell Interference Detection and Mitigation in a Multiple Networking Environment 71
R. Ravi, R. Mallika Pandeeswari, R. Kabilan, and S. Shargunam
6.1 Introduction 72
6.2 Proposed System 74
6.3 OFDMA and SCFDMA 74
6.4 Results and Discussion 74
6.4.1 BER–SNR graph of two users. 75
6.4.2 BLER–SNR graph of two user 75
6.4.3 SE-SNR graph. 78
6.5 Comparison of Detector Performance as a Result of Shot Interference 78
6.6 BER–SNR Graph of Different Detectors 78
6.7 Channel MSE–ESN0 Graph. 80
6.8 Conclusion. 80
References 81

Chapter 7
A Baseband Transceiver for MIMO-OFDMA in Spatial Multiplexing Using Modified V-BLAST Algorithm 83
R. Kabilan, R. Ravi, S. Shargunam, and R. Mallika Pandeeswari
7.1 Introduction 84
7.1.1 OFDM modulation 84
7.1.2 FDMA. 84
7.1.3 OFDMA. 85
7.1.4 MIMO OFDM. 85

7.2 Existing Method . . . 85
7.2.1 Synchronization algorithms for MIMO OFDMA systems . . . 86
7.3 MIMO Transceiver . . . 86
7.4 Proposed Method . . . 86
7.4.1 Module description . . . 87
7.4.2 Proposed modified V-BLAST algorithm . . . 88
7.5 Result and Discussion . . . 88
7.6 Conclusion . . . 92
References . . . 93

Chapter 8
Hardware Implementation of OFDM Transceiver Using Simulink Blocks for MIMO Systems **95**
R. Ravi, J. Zahariya Gabriel, R. Kabilan, and R. Mallika Pandeeswari
8.1 Introduction . . . 96
8.2 Existing System . . . 97
8.2.1 Fast ICA . . . 97
8.2.2 Efficient variant of fast ICA algorithm (EFICA) . . . 97
8.2.3 Sphere decoding algorithm . . . 97
8.3 Proposed System . . . 98
8.4 MIMO-OFDM . . . 98
8.5 Channel Estimation (CE) . . . 98
8.6 Flow Diagram . . . 98
8.6.1 Input sample . . . 98
8.6.2 Serial to parallel converter . . . 99
8.6.3 AWGN channel . . . 99
8.6.4 Mapper . . . 99
8.6.5 FFT block . . . 99
8.6.6 IFFT block . . . 100
8.6.7 BER . . . 100
8.7 Module Explanation . . . 100
8.7.1 OFDM modulation/demodulation . . . 100
8.7.2 FFT/IFFT block . . . 100
8.7.3 OFDM transmitter . . . 100
8.7.4 OFDM receiver . . . 101
8.8 Results and Discussion . . . 101
8.8.1 Selection of voice source . . . 101
8.8.2 MIMO block design process . . . 102

8.8.3 Synthesis process 102
8.8.4 RTL schematic. 104
8.8.5 Technology schematic 105
8.8.6 Power estimation 106
8.8.7 Static power 107
8.8.8 Power estimation 107
8.9 Conclusion and Future Work 108
References 108

Chapter 9
Empowering Radio Resource Allocation to Multicast Transmission System Using Low Complexity Algorithm in OFDM System **111**
R. Kabilan, R. Ravi, J. Zahariya Gabriel, and R. Mallika Pandeeswari
9.1 Introduction 112
9.2 Existing System 112
9.2.1 Conventional multicast scheme 113
9.2.2 Radio resource management (RRM) algorithm 113
9.3 Proposed System 113
9.4 Multi Rate Scheme 113
9.4.1 OFDMA framework 113
9.4.2 Utilisation of resources depending upon subgroups. . 114
9.4.3 Channel state information (CSI). 114
9.4.4 Signal to interference plus noise ratio (SINR) 114
9.5 Frequency Domain Subgroup Algorithm (FAST) 115
9.6 Results and Discussions 115
9.6.1 Separate cell's creation 115
9.6.2 After every round, the average energy in each user . . 116
9.6.3 Power allocation to each and every subgroup 116
9.6.4 Capacity allocation and allocation of LAMDA 116
9.6.5 The performances estimation of various parameters . 118
9.7 Conclusion. 118
References 119

Chapter 10
Survey on RF Coils for MRI Diagnosis System **123**
K. Sakthisudhan, N. Saranraj, and C. Ezhilazhagan
10.1 Introduction 123
10.2 Survey of Literature 125
10.2.1 Design of transceiver RF coils 125

10.2.2 The development of RF based MRI coils 126
10.2.3 Research development of industry version of MRI coils 129
10.3 Proposed Methodology 130
10.3.1 A design a coils by meta-materials 130
10.3.2 Implementation using big data digitization analysis through wireless networks 132
10.3.3 To design a flexible adaptive multituned RF coils. . . 133
10.4 Conclusion. 134
References 135

Chapter 11
Wireless Sensing Based Solar Tracking System Using Machine Learning **139**
S. Saroja, R. Madavan, Jeevika Alagan, V.S. Chandrika, and Alagar Karthick
11.1 Introduction 140
11.1.1 Purpose 140
11.1.2 Overview 141
11.2 Solar Tracking System 141
11.2.1 Architectural description 141
11.2.2 Main components 142
11.2.2.1 Servomotor 142
11.2.2.2 LDR 143
11.2.2.3 Solar panel. 144
11.2.4 Limitations. 145
11.2.5 Dependencies and assumptions 145
11.2.6 Specifications for requirements 145
11.2.6.1 Requirement for external interface 146
11.2.6.2 Requirements, both functional and non-functional 146
11.3 Machine Learning Algorithms 147
11.3.1 Supervised learning 148
11.3.1.1 Classification 148
11.3.1.2 Regression. 150
11.4 Machine Learning Algorithm for Solar Tracking System. . . 151
11.4.1 SVM for solar tracking system 151
11.4.1.1 Steps in python implementation of SVM . . 152
11.4.2 Linear regression for solar tracking system 153
11.4.2.1 Steps in python implementation of linear regression 153

11.5 Implementation 154
11.6 Conclusion. 158
References 158

Chapter 12
Gain and Bandwidth Enhancement of Pentagon Shaped Dual Layer Parasitic Microstrip Patch Antenna for WLAN Applications **163**
Ambavaram Pratap Reddy, and Pachiyaannan Muthusamy
12.1 Introduction 164
12.2 Pentagon Single Layer Design 165
12.3 Pentagon Dual Layer Design 166
12.4 Analysis of the Dual Layer Pentagon with Two Parasitic Elements 166
12.5 Analysis of the Dual Layer Pentagon with Two Parasitic Elements 169
12.6 Conclusion. 171
References 173

Chapter 13
Quantum Cascade Lasers – Device Modelling and Applications **175**
M. Ramkumar, Pon Bharathi, A. Nandhakumar, P. Srinivasan, D. Vedha Vinodha, and P. Ashok
13.1 Introduction 175
13.2 Non-Linear Frequency Generation 176
13.3 QCL Based Interferometry 176
13.4 Frequency Instabilities in THz QCLs 177
13.5 Design Optimization of Cavity in QCLs 178
13.6 THz QCLs Based on HgCdTe Material Systems 179
13.7 Optical Beam Characteristics of QCL. 179
13.8 Temperature Degradation in THz QCLs 180
13.9 Impedance Characteristics of QCLs. 180
13.10 Free Space Optical Communication Using QCLs. 181
13.11 Free Space Optical Communication Using Temperature Dependent QCLs 182
13.12 Conclusion. 182
References 183

Chapter 14
Design of Broad band Stacked Fractal Antenna with Defective Ground Structure for 5G Communications **185**
G. Rajyalakshmi, and Y. Ravi Kumar
14.1 Introduction 185
14.2 Introduction to the Fractal Concept 187
14.2.1 The fractal geometry 187
14.3 Antenna Design with a Single Layer Fractal 187
14.4 Antenna Design with a Dual Layer Fractal 188
14.5 Conclusion 190
14.6 Acknowledgement 193
References 193

Chapter 15
Performance Analysis of T Shaped Structure for Satellite Communication **197**
Jacob Abraham, R. S. Kannadhasan, R. Nagarajan, and Kanagaraj Venusamy
15.1 Introduction 198
15.2 T Shaped Structure Antenna 198
15.3 Results and Discussion 200
15.4 Conclusion 202
References 203

Index **205**

About the Editors **207**

Preface

This book provides an overview of Antennas and Wireless Communication. The Smart Antennas for Wireless Communications together top academic scientists, researchers, and research scholars to discuss and share their experiences and research findings on all areas of Smart Antennas, Electromagnetic Interference, and Microwave Antennas for Wireless Communications. Smart Antennas or Adaptive Antennas are multi-antenna components on one or both sides of a radio communication connection, combined with advanced signal processing algorithms. They've evolved into a critical technology for third-generation and beyond mobile communication systems to meet their lofty capacity and performance targets. It seems that a significant capacity gain is achievable, particularly if they are employed on both sides of the connection. There are several essential characteristics of these systems that need scientific and technical investigation. The International Workshop on Smart Antennas and Microwave Antennas is an annual gathering that offers a forum for researchers to share their newest findings. Its goal is to include both theoretical and technological aspects of intelligent antennas, such as in mobile communications and localization. Beamforming, massive MIMO, network MIMO, mmWave transmission, compressive sensing, MIMO radar, sensor networks, vehicle-to-vehicle communication, location, and machine learning were among the subjects covered in this year's program. The various discussions on a variety of current hot topics, including 5G propagation, MIMO, and array antennas, Optical nano-antennas, Scattering and diffraction, Computational electromagnetics, Radar systems, Plasmonics and nanophotonics, and Advanced EM materials and structures like metamaterials and metasurfaces. We would like to take this opportunity to thank our family members and friends, encouraged us a lot during the preparation of this book. First and most obviously, we give all the glory and honor to our almighty Lord for his abundant grace that sustained me for the successful completion of this book. I would like to thank the authors for their contribution to this edited book. I would also like to thank River Publisher and its whole team for facilitating the work and providing me the opportunity to be a part of this work.

List of Contributors

A. Nandhakumar, *Department of ECE, Dhaanish Ahmed Institute of Technology, Coimbatore*

Alagar Karthick, *Associate Professor, Renewable Energy Lab, Department of Electrical and Electronics Engineering, Coimbatore-641404, Tamilnadu, India*

Ambavaram Pratap Reddy, *Advanced RF Microwave & Wireless Communication Laboratory, Vignan's Foundation for Science Technology and Research (Deemed to be University), Vadlamudi, Andhra Pradesh, India*

C. Ezhilazhagan, *Assistant Professor, Dr. N.G.P. Institute of Technology, Coimbatore, Tamilnadu, India*

D. Vedha Vinodha, *Department of ECE, JCT college of Engineering and Technology, Coimbatore*

Dr. Y. Ravi Kumar, *Scintist-G (Retd) DLRL, DRDO, Hyderabad, Telengana, India*

G. Rajyalakshmi, *University College of Engineering, Osmania University, Hyderabad, Telengana, India*

J. Zahariya Gabriel, *Associate Professor, Department of ECE, Francis Xavier Engineering College, Tirunelveli, India*

Jacob Abraham, *Associate Professor, Department of Electronics, B P C College, Piravom, Eranakulam, Kerala, India*

Jeevika Alagan, *Assistant professor, PG and Research department of chemistry, Thiagarajar college, Madurai-625009, Tamilnadu, India*

K. Sakthisudhan, *Professor, Dr. N.G.P. Institute of Technology, Coimbatore, Tamilnadu, India*

Kanagaraj Venusamy, *Control Systems Instructor, Department of Engineering, University of Technology and Applied Sciences-AI Mussanah, AI Muladdha*

M. Ramkumar, *Associate Professor, Department of ECE, Sri Krishna College of Engineering and Technology, Coimbatore*

N. Saranraj, *Research Scholar, Anna University, Chennai, Tamilnadu, India*

P. Srinivasan, *Department of ECE, Amrita College of Engineering and Technology, Nagercoil*

Pachiyaannan Muthusamy, *Advanced RF Microwave & Wireless Communication Laboratory, Vignan's Foundation for Science Technology and Research (Deemed to be University), Vadlamudi, Andhra Pradesh, India*

Pon Bharathi, *Assistant Professor, Department of ECE, Amrita College of Engineering and Technology, Nagercoil*

R. Kabilan, *Associate Professor, Department of ECE, Francis Xavier Engineering College, Tirunelveli, India*

R. Madavan, *Associate Professor & Head Department of Electrical and Electronics Engineering, PSR Engineering College, Sivakasi, Tamilnadu, India*

R. Mallika Pandeeswari, *Full Time Ph.D Scholar, Dept of ECE, Francis Xavier Engineering College, Tirunelveli, India*

R. Nagarajan, *Professor, Department of Electrical and Electronics Engineering, Gnanamani College of Technology, Tamilnadu, India*

R. Ravi, *Professor, Department of CSE, Francis Xavier Engineering College, Tirunelveli, India*

R S. Kannadhasan, *Assistant Professor, Department of Electronics and Communication Engineering, Cheran College of Engineering, Tamilnadu, India*

S. Saroja, *Assistant Professor/ Information Technology, Mepco Schlenk Engineering College, Sivakasi, Tamilnadu, India*

S. Shargunam, *Full Time Ph.D Scholar, Dept of ECE, Francis Xavier Engineering College, Tirunelveli, India*

V.S. Chandrika, *Associate Professor, Renewable Energy Lab, Department of Electrical and Electronics Engineering, Coimbatore-641404, Tamilnadu, India*

List of Figures

Figure 1.1	Steps in fast ICA algorithm.	4
Figure 1.2	Modules of the proposed method.	4
Figure 1.3	Signal representation of the man voice..	5
Figure 1.4	Signal representation of the music.	6
Figure 1.5	Signal representation of the mixed signal1..	6
Figure 1.6	Signal representation of the mixed signal3..	7
Figure 1.7	Signal representation of the extracted man voice.	7
Figure 1.8	Signal representation of the extracted music..	8
Figure 1.9	Signal representation of the separated noise signal 1.	8
Figure 1.10	Signal representation of the separated noise signal 3.	9
Figure 2.1	Block diagram of proposed method.	16
Figure 2.2	Detailed proposed system.	17
Figure 2.3	OFDM channel estimation.	19
Figure 2.4	CIR of the wireless channel.	20
Figure 2.5	SNR Vs PER of MIMO OFDM for 2X2 users..	21
Figure 2.6	BER Vs SNR of fiber-wireless uplink.	21
Figure 2.7	BER Vs SNR for MIMO with 16 QAM.	22
Figure 2.8	QAM modulated signal.	23
Figure 2.9	Convergence graph..	24
Figure 3.1	Components of communication system.	30
Figure 3.2	Block diagram for proposed system.	31
Figure 3.3	Process flow chart.	33
Figure 3.4	Command window.	34
Figure 3.6	Node 1 starts transmitting data..	35
Figure 3.5	Initialization of nodes.	35
Figure 3.8	Transmission of data through relay node..	36
Figure 3.7	Choosing the shortest path.	36
Figure 3.9	Node 8 starts transmitting data..	37
Figure 3.11	Transimision of data from node 7 to node 6.	38
Figure 3.10	Loss of packets..	38
Figure 3.12	Transmision of data from node 2 to node 4.	39
Figure 3.13	Transmission of data bidirectionally.	39

Figure 3.14 Completion of transmission from node 8 to 0. 40
Figure 3.15 Coverage provided by dynamic base station. 40
Figure 3.16 Retransmission of dropped packets. 41
Figure 3.17 Reception of acknowledgement. 42
Figure 3.18(a) Graph for lifetime. 42
Figure 3.18(b) Lifetime vs node ready. 43
Figure 3.19(a) Graph for output. 43
Figure 3.19(b) Graph for average throughput. 44
Figure 4.1 Relay network with M relays, source, and destination. 50
Figure 4.2 Skeletal view of proposed method. 50
Figure 4.3 SNR vs BER. 52
Figure 4.4 Shows that the relation between the SNR and BER between various antenna connected to relays. 52
Figure 4.5 Shows that M = 1, L = 2 relay signal level where every node got connected with two antennas and obtain the minimum bit error rate. 53
Figure 4.6 Shows that the relationship between SNR and BER when different number of relays are used. 53
Figure 4.7 Shows that total relay transmit power limits. 54
Figure 4.8 Shows that the proposed SINR method Max-Min Level.. 54
Figure 4.9 It shows how increasing the transmit power of both sources enhances the suitability of both the optimization processes and decreases the bandgap between them.. 55
Figure 4.10 Indicates that for long-distance broadcasting, the smallest possible power is necessary.. 55
Figure 4.11 Illustrates that graphing which shows the require SINR for various source transmit power Ps yields vs total power consumption the desired result. 56
Figure 5.1 Flow chart for VB decoding algorithm.. 61
Figure 5.2 Flow chart for SE decoding algorithm. 63
Figure 5.3 FPGA-based SoC architecture for sphere decoder implementations. 64
Figure 5.4 Simulated output of VB. 65
Figure 5.5 Test bench output of VB. 65
Figure 5.6 Simulated output of SE.. 66
Figure 5.7 Test bench of output SE. 66
Figure 5.8 Synthesis report of VB. 67

Figure 5.9 Synthesis report of S. 67
Figure 6.1 Wireless cellular networks [from internet]. 72
Figure 6.2 Femtocell. 73
Figure 6.3 SCFDMA and OFDMA transmission and reception.. 75
Figure 6.4 BER–SNR graph of two users. 76
Figure 6.5 BLER–SNR graph of two uses.. 77
Figure 6.6 E-SNR graph of two users. 79
Figure 6.7 BER–SNR Graph.. 79
Figure 6.8 Channel MSE–ESN0 Graph. 80
Figure 7.1 OFDM Modulation subcarrier allocation diagram [from Internet]. 84
Figure 7.2 Carrier allocation for OFDMA signal [from Internet]. 85
Figure 7.3 Block diagram of transceiver.. 87
Figure 7.4 Propose algorithm for modified V BLAST.. 89
Figure 7.5 ML estimation. 90
Figure 7.6 BER vs Eb/No. 90
Figure 7.7 BER vs Average Eb/No.. 91
Figure 7.8 BER vs SNR. 91
Figure 7.9 MSE vs SNR. 92
Figure 8.1 MIMO systems [from internet].. 96
Figure 8.2 Proposed flow diagram.. 99
Figure 8.3 Input audio wave. 101
Figure 8.4 Limited input wave.. 102
Figure 8.5 MIMO block design process. 103
Figure 8.6 Synthesis process.. 103
Figure 8.7 RTL schematic. 104
Figure 8.8 Technology schematic. 105
Figure 8.9 Total on chip power.. 106
Figure 8.10 Static power. 107
Figure 9.1 Blocks involved in multicast sub grouping system.. . 114
Figure 9.2 Creation of each cell in multicast sub grouping technique. 115
Figure 9.3 Average energy in each user Vs Round number. . . . 116
Figure 9.4 Amount of resources allocated.. 117
Figure 9.5 Allocation of capacity and wavelength for each subgroup. 118
Figure 9.6 Comparison of various parameters vs number of operations needed to convergence. 119
Figure 10.1 Flexible anterior RF receiver coil with 64-element array. 127

Figure 10.2 Left side of the figure shows an MRI scan coil separated from the patient and the right side of the figure shows that the coils are positioned against the patient with a significantly improved quality.. 130
Figure 11.1 Solar panel assembly. 141
Figure 11.2 Main components of solar tracking system. 142
Figure 11.3 SVM classification. 152
Figure 12.1 Single layer pentagon Antenna.. 165
Figure 12.2 Proposed results (a) S_{11}-Parameter (b) VSWR. 165
Figure 12.3 Proposed antennas Gain at 5.3GHz. 166
Figure 12.4 Proposed dual layer Antenna.. 167
Figure 12.5 Proposed antenna results (a) S_{11}-Parameter (b) VSWR (c) Gain.. 167
Figure 12.6 Proposed antenna dual layers with two parasitic element. 168
Figure 12.7 Proposed antenna results (a) S_{11}-Parameter (b) VSWR (c) Gain.. 168
Figure 12.8 Proposed antenna dual layers with four parasitic elements. 169
Figure 12.9 Proposed antenna results (a) S_{11}-Parameter (b) VSWR (c) Gain.. 170
Figure 12.10 Surface current distributions at 5.4GHz. 170
Figure 12.11 Radiation pattern at 5.4GHz results and discussion. . 171
Figure 12.12 Simulated and measured results. 171
Figure 12.13 Prototype proposed antenna with measurement setup. 172
Figure 13.1 Model of a pulsed QCL under optical feedback. . . . 177
Figure 13.2 Experimental setup for the 4.7-THz experiments. . . 178
Figure 13.3 Block diagram of a generic FSO Link. 181
Figure 14.1 Design of the fractal geometries; (a) zero iteration, (b) First iteration, (c) second iteration. 188
Figure 14.2 Dimensions of the single layer proposed antenna. . . 189
Figure 14.3 S-Parameter results a-c iterations (d) surface current distribution at 3.5GHz. 189
Figure 14.4 (a) Iteration-Zero gain (b) Iteration-1 gain (c) Iteration-2gain. 190
Figure 14.5 (a) Dual layer fractal front view (b) DGS back view (c) feed line (d) dual layer structure side view. 191
Figure 14.6 (a) S-Parameter result (b) radiation pattern at 3.5GHz.. 192

Figure 14.7 (a) Surface current distribution at 3.5GHz (b) Gain at 3.5GHz. 192
Figure 14.8 Gain of the single and dual layer. 192
Figure 15.1 T Shaped structure. 199
Figure 15.2 Return loss of the antenna. 201
Figure 15.3 VSWR of the antenna. 201
Figure 15.4 Gain of the antenna.. 202
Figure 15.5 Radiation pattern of the antenna. 203

List of Tables

Table 5.1 Experimental result.. 68
Table 8.1 Comparison table.. 108
Table 12.1 Pentagon antenna dimensions. 166
Table 12.2 Comparison analyses.. 171
Table 12.3 Comparisons of present and previous works.. 172
Table 14.1 Compares the proposed work to the existing work.. . 193

List of Abbreviations

IVP	Initial value problems
BVP	Boundary value problems
SLP	Sturm liouville problem
RHS	Right hand Side
ADM	Adomian decomposition method
HAM	Homotopy analysis method
HPM	Homotopy perturbation method
DTM	Differential transform Method
R-K	Runge-Kutta

Chapter 1

Speech Signal Extraction from Transmitted Signal Using Multilevel Mixed Signal

R. Kabilan[1], and R. Ravi[2]

[1]Associate Professor, Department of ECE, Francis Xavier Engineering College, Tirunelveli, India
[2]Professor, Department of CSE, Francis Xavier Engineering College, Tirunelveli, India
Email: rkabilan13@gmail.com; fxhodcse@gmail.com

Abstract

Speech signal extraction is becoming increasingly significant in a range of applications, including mobile phones, conferencing equipment, and other similar devices. This project shows how to separate voice sounds in a noisy environment using a blind method. It is a mechanism for isolating an independent signal from a mixed signal that saves energy. To separate speech signals, the proposed technique uses Fast Independent Component Analysis (FastICA). The study looks into the effects of the environment, like noise, the number of locations, and their sources, on the method's performance.

1.1 Introduction

When we imagine ourselves at a party, our ears pick up a variety of sounds: a friend's voice, other people's voices, background music, ringing phones, and so on. If you focus, you can hear what someone is saying while filtering out all other noises. The cocktail party effect is the capacity to focus and pinpoint a single source [12]. When we record sounds in various locations in the room, the playback is a chaotic mixture of various sounds. so that we can hear only some words and not be able to hear the conversation. Blind source separation algorithms can extract and separate each discourse if there are as many microphones in the room as people. As a result, we'd be able to hear everything in the room as a result of this. BSS is a method for separating the original sources from an array of transducers or sensors without getting the real source. Blind source separation is a normal signal processing method for extracting statistically independent source signals from linear mixes with little or no knowledge of the sources or mixing settings [13].

Principal Component Analysis is mostly used in some tasks in challenging situations. The main differences between the two methods are that the former uses non-Gaussian and independent sources, while the latter uses uncorrelated sources with Gaussian distributions [14]. For instance, in a surveillance application, when the goal is to detect a certain voice among a large number of others,

Furthermore, BSS is involved in wireless transmission to reduce co-channel interference in situations where multiple antennas are used. The BSS method is also used in various fields of biomedical signal processing techniques. It is also used in ECG measurement. Separation of the mother's ECG from the fetal ECG is one of the uses of FECG [11]. Separating the MECG from a twin fetus is also a complicated circumstance. In the fields of digital signal processing and neural networks, field-programmable gate array (FPGA) technology has recently been the preferred method of implementation [10]. The majority of digital signal processing techniques have been implemented using FPGA. Without any fabrication delays, DSP algorithms can be designed, tested, and implemented on an FPGA chip. FPGAs are made up of the following components:

These resources are ideal for DSP applications because they offer great performance and low power consumption. A hardware description language (HDL) such as VHDL or Verilog can be used to specify an FPGA's behavior, or a schematic-oriented design tool can be used to arrange blocks of existing functionality [3]. The FPGA can download a bit file created during the compilation and synthesis process. FPGAs are comparable to custom ASIC designs in that they can construct and test proposed designs rather than send them to a manufacturer and wait for the chip to be tested [1].

The implementation of the ICA technique for higher-order data inputs is the thesis's key challenge. The circuit's complexity increases exponentially in proportion to the size of the demixing matrix W. In addition, the method is implemented using an FPGA.

1.2 Literature Survey

1.2.1 The fast ICA algorithm revisited: Convergence analysis

One of the most prominent approaches to solving problems in ICA is the fast ICA algorithm. Only for the one unit scenario, in which one of the rows of the separation matrix is evaluated, has a convergence analysis been presented.

It is used to increase processing speed [9]. The noise-free system was the reason for the introduction of fast ICA. The goal is to figure out what the unknown mixing matrix "A" is and what the unknown independent sources "S" is. First, the signal sources are transformed into a noise-free signal. Fast ICA is a never-ending process.

1.2.2 FPGA implementation of IC algorithm for blind signal separation and noise cancelling

Implementation of FPGA In real-time, ICA reported on blind source separation and noise cancellation. It used an ICA-based method with a multipath special digital processor to give massive compute capability. Speech recognition, as a generic method, performs poorly [2]. The noise cancellation method can be used to remove noise by using a reference signal. For noise cancellation, the LMS technique is often used. However, Independent Component Analysis outperforms Least Mean Square in FPGA implementations that leverage the modular design idea.

1.2.3 Subjective comparison and evaluation of speech enhancement algorithms

This article details the creation of a noisy speech corpus that may be used to test speech enhancement techniques. This corpus is used to assess 14 speech enhancement approaches, which are divided into four categories: spectrum subtractive, subspace, statistical-model based, and general algorithms [8]. The findings of the subjective tests are presented in this publication. The noise-estimation approach did not significantly improve the performance of the log-MMSE algorithm in this case. The log-MMSE algorithm's performance was not improved by including speech-presence uncertainty [4].

1.3 Proposed Systems

Up to four input sensors can be used in the proposed BSS concept. Because the model can differentiate up to four mixed signals in the mixture, this assumption is broken. The sources in the mix are assumed to be non-Gaussian and independent.

1.3.1 FASTICA using symmetric orthogonalization

This thesis focuses on using symmetric orthogonalization to implement the Fast ICA method. Alternative strategies have been suggested. One method to speed up the process of symmetrical orthogonalization is to employ an

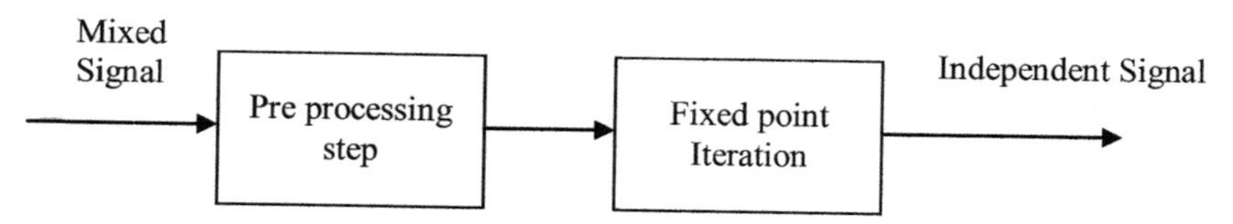

Figure 1.1 Steps in fast ICA algorithm.

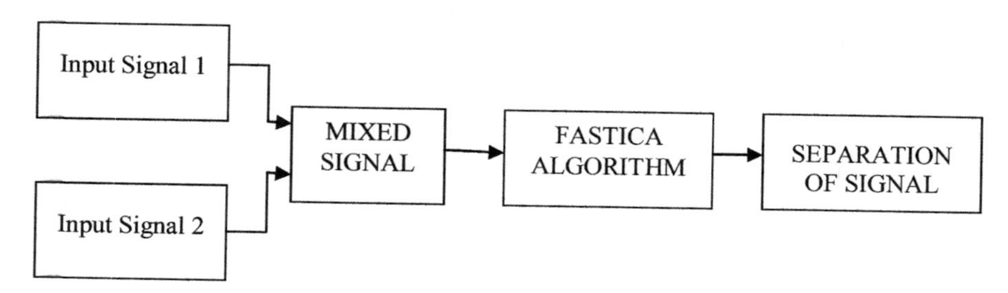

Figure 1.2 Modules of the proposed method.

iterative model. The iterative strategy is used in the simulation shown in the picture to separate the sources. The x-axis depicts the number of iterations required for convergence, while the y-axis represents the difference in error between the current iterative result and the final algebraic result. When the two answers match, the simulation concludes with a zero error.

Because we are just modelling the number of iterations and not the time required to obtain a solution, the figure does not imply that the iterative solution is slower than the algebraic one. In fact, the iterative method's 15 repetitions converge to a solution significantly faster than the algebraic method's computations [5].

1.3.2 FPGA implementation

The FastICA algorithm is implemented in an FPGA using the XILINX virtex5-XC5VLX50t FPGA device, as is well known. The LX50t processor outperforms the other Virtex 5 chips in terms of speed and area. The VHDL language is used to implement the design. Because the system is built to handle higher-order data (four sensors), hierarchy is used throughout the design to provide the designer with more control over the overall hardware structure and to keep track of each block's overflow and underflow [6,7].

Furthermore, DSP systems based on floating-point arithmetic necessitate a large amount of hardware and might result in inefficient design, particularly for FPGA implementation. Fixed-point representation, on the other hand,

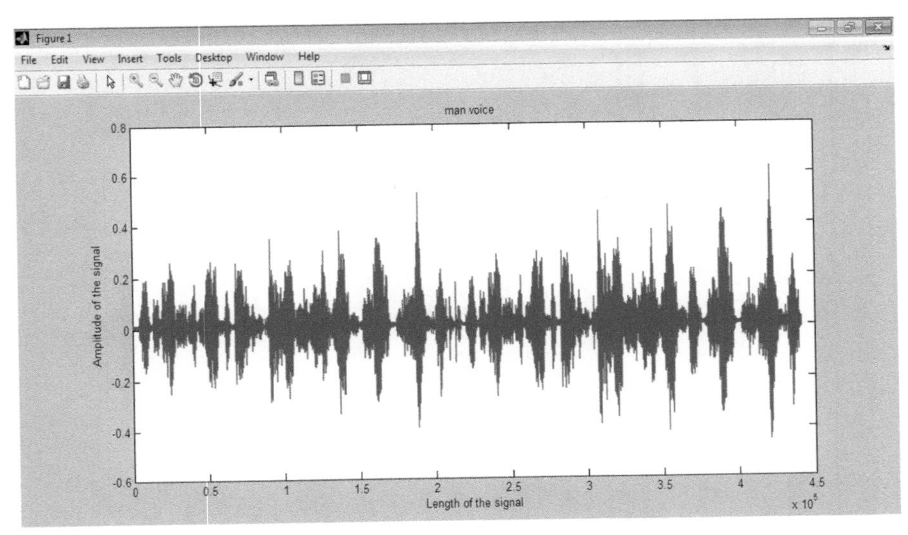

Figure 1.3 Signal representation of the man voice.

allows for more efficient hardware design. Two's complement fixed-point arithmetic is employed in this thesis. Several simulations were used to determine the word length. When a small word length was employed, the majority of the findings were incorrect because the small word length was insufficient to represent the values. The word length for the centering and covariance blocks was chosen to be 16 bits since the calculations for the centering and covariance were not complicated, and 16 bits were sufficient to represent intermediate variables such as signals and storage elements inside the system.

1.4 Results and Discussion

Output Waveform

Figure 1.3 shows the signal representation of the man's voice, which is represented as the length of the signal in the x-axis and the amplitude of the signal in the y-axis.

Figure 1.4 shows the signal representation of the music, which is represented as the length of the signal on the x-axis and the amplitude of the signal on the y-axis.

Figure 1.5 shows the signal representation of the mixed signal, which is the mixing of the man's voice with the second signal and the random matrix value. The length of the signal is denoted on the x-axis and the amplitude of the signal on the y-axis.

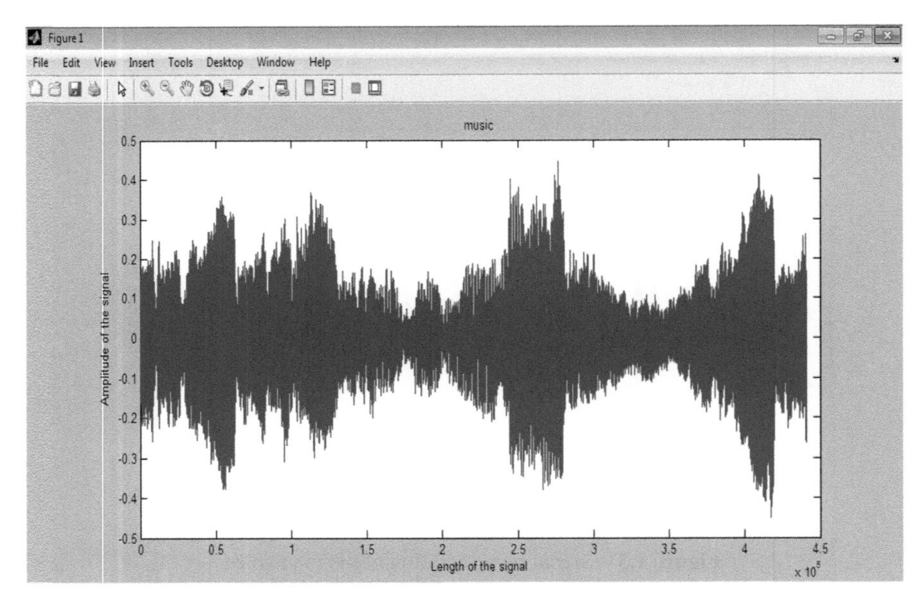

Figure 1.4 Signal representation of the music.

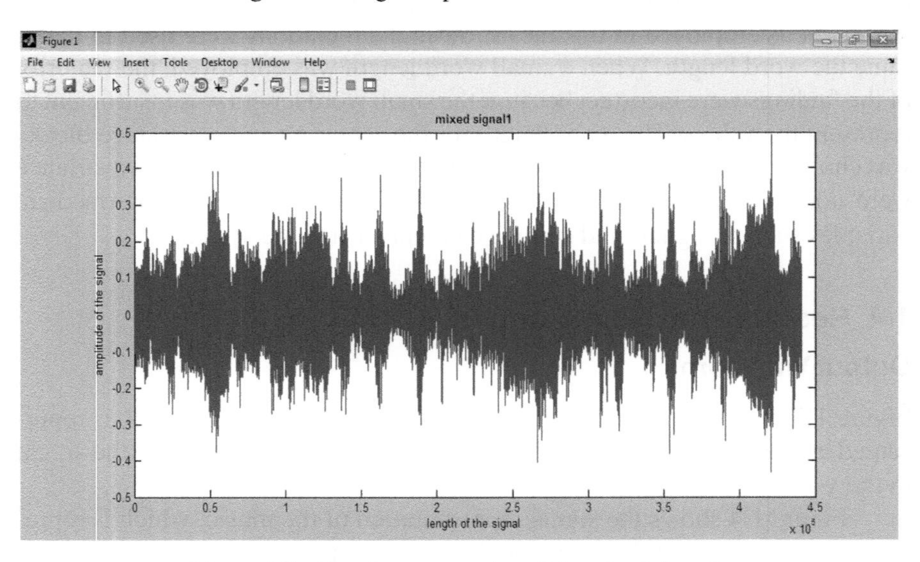

Figure 1.5 Signal representation of the mixed signal1.

Figure 1.6 shows the signal representation of the mixed signal3, which is the mixing of the music with the first signal and the random matrix value. The length of the signal is denoted on the x-axis and the amplitude of the signal on the y-axis.

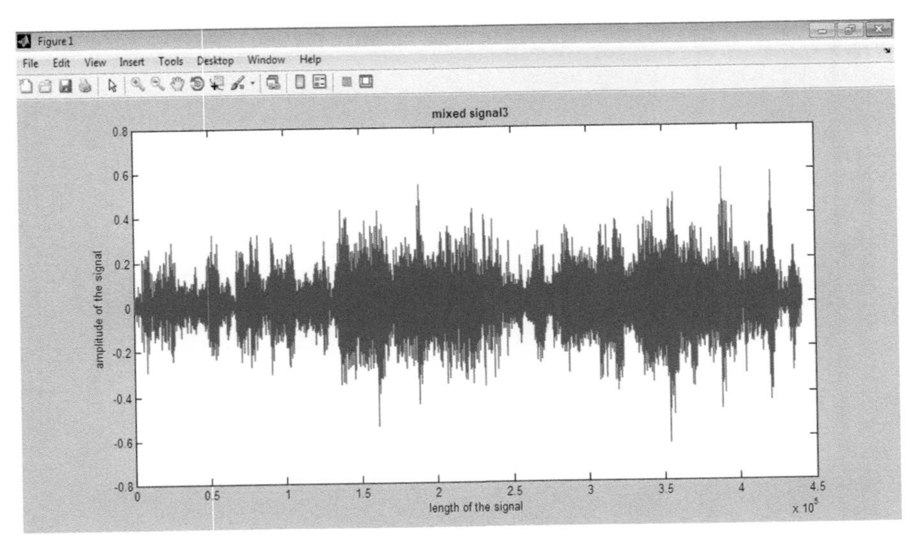

Figure 1.6 Signal representation of the mixed signal3.

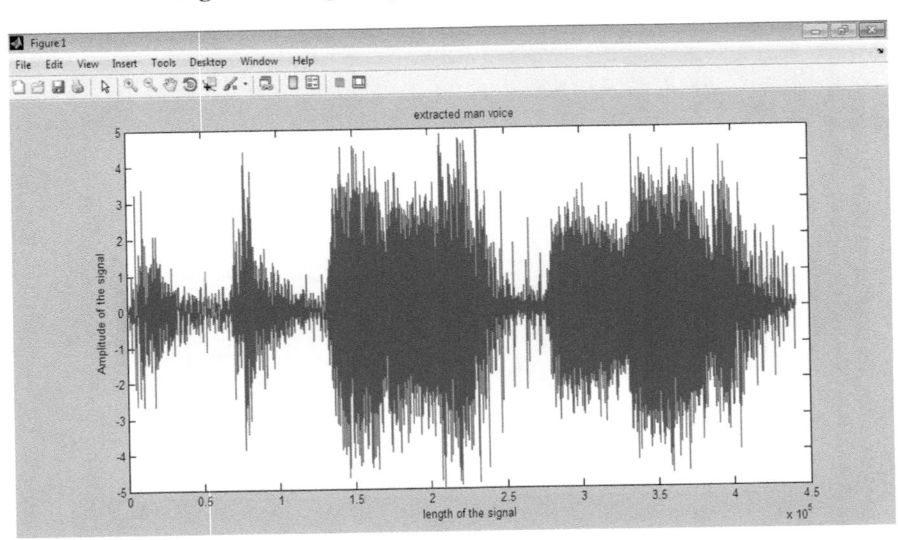

Figure 1.7 Signal representation of the extracted man voice.

Figure 1.7 represents the extracted man voice from the mixed signal (a combination of the man voice with the second signal and with the random matrix values). The length of the signal is denoted on the x-axis and the amplitude of the signal on the y-axis.

Figure 1.8 represents the extracted music from the mixed signal 3 (a combination of the music with the first signal and with the random matrix

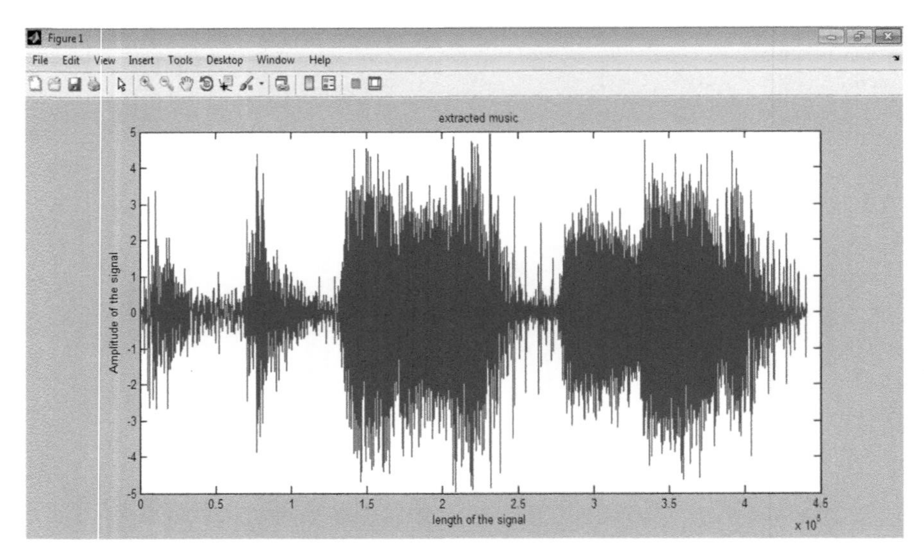

Figure 1.8 Signal representation of the extracted music.

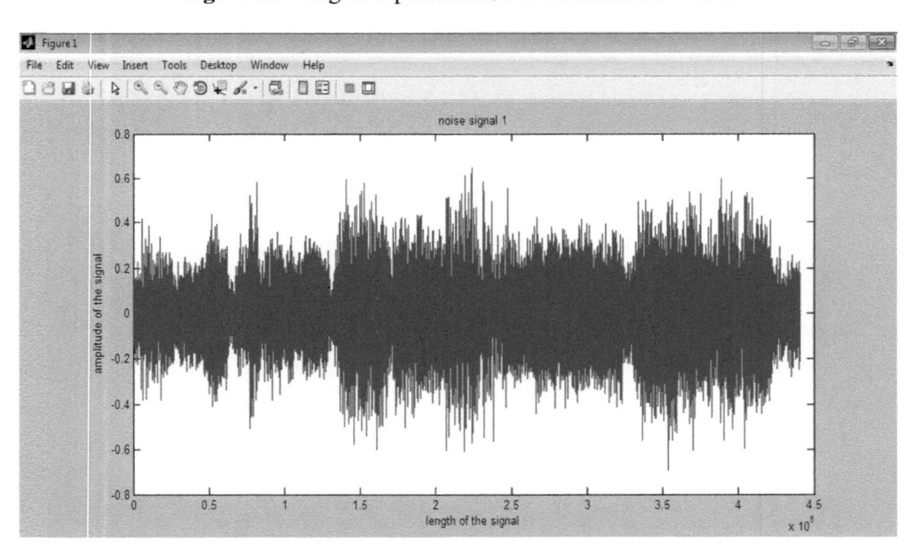

Figure 1.9 Signal representation of the separated noise signal 1.

values). The length of the signal is denoted on the x-axis and the amplitude of the signal on the y-axis.

Figure 1.9 represents the separated from the mixed signal1 (combination of the man's voice with the second signal and with the random matrix values). The length of the signal is denoted on the x-axis and the amplitude of the signal on the y-axis.

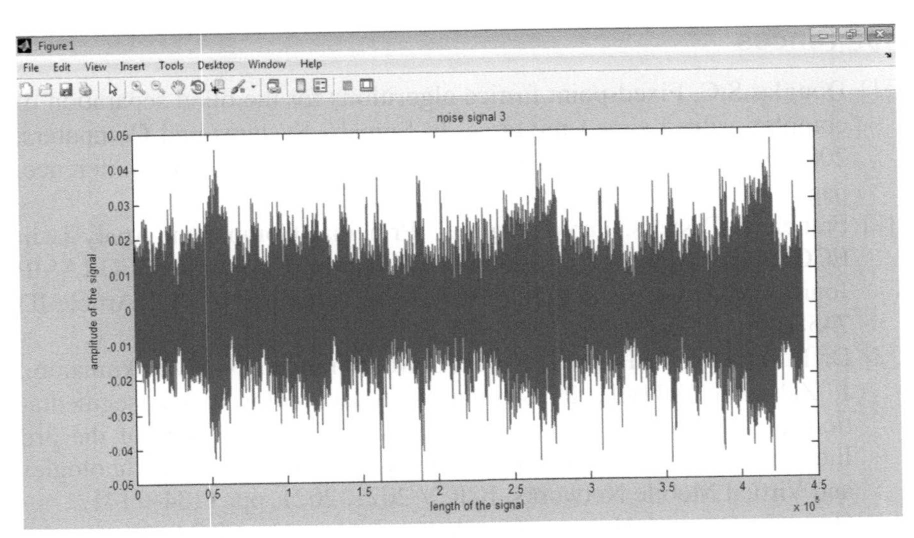

Figure 1.10 Signal representation of the separated noise signal 3.

Figure 1.10 represents the separated noise signal 3 from the mixed signal 1 (a combination of the music with the first signal and with the random matrix values). The length of the signal is denoted on the x-axis and the amplitude of the signal on the y-axis.

1.5 Conclusion

We presented the FastICA architecture in this project, which is energy-efficient, low-cost, and has a short calculation time. It looks into the effect of the environment on FastICA's speech separation ability. We concentrated on noise, the kind of noise, the number of sources, and the mixing ratio. We used MATLAB to develop the FastICA script, and the function Kurtosis was chosen as the measure of independence. We used the SDR, SIR, and SAR parameters to test the results of this script on 10 second-long voice samples. By adding an analog to a digital converter before the whitening block to gather data and a digital to analogue converter at the output to turn the output back to the analogue, the proposed architecture can be utilized as a building block to separate real-time signals. In the future, we'd like to compare our results to those of other ways (techniques) of separating mixed signals. Finally, this research can be applied to signal separation in various applications such as cardiac electrical imaging, data mining, and wireless communication.

References

[1] Douglas.S.C, Fixed-point fastica algorithms for the blind separation of complex valued signal mixtures. In Signals, Systems and Computers, 2005. Conference Record of the Thirty-Ninth Asilomar Conference, pages 1320– 1325, Pacific Grove, CA, 2005.

[2] Francisco Castells, Pablo Laguna," Principal Component Analysis in ECG Signal Processing "Hindawi Publishing Corporation EURASIP Journal on Advances in Signal Processing Volume Sep 2007, Article ID 74580.

[3] Dr. R. Kabilan., G. Prince Devaraj., U. Muthuraman., N. Muthukumaran., J. Zahariya Gabriel., R. Swetha, 'Efficient color image segmentation using fastmap algorithm', IEEE Xplore, Proceedings of the 3rd International Conference on Intelligent Communication Technologies and Virtual Mobile Networks, ICICV 2021, 2021, pp. 1134–1141.

[4] Hongtao Du and Hairong Qi "A Reconfigurable FPGA System for Parallel Independent Component Analysis". EURASIP Journal on Embedded Systems. volume 2006, Article ID 23025, Pages 1–12

[5] J. Zahariya Gabriel., Dr. R. Kabilan., G. Prince Devaraj., U. Muthuraman, 'Facial authentication system by combining of feature extraction using raspberry Pi', IEEE Xplore, Proceedings of the 3rd International Conference on Intelligent Communication Technologies and Virtual Mobile Networks, ICICV 2021, 2021, pp. 1142–1145.

[6] Kevin Banovi´c , "Blind Adaptive Equalization for QAM Signals: New Algorithms and FPGA Implementation", University of Windsor, Master's Thesis, 2006

[7] Dr. U. Muthuraman, Dr. R. Kabilan, 'A high power EV charger based on modified bridgeless LUO converter for electric vehicle', IEEE Xplore, Proceedings of the 3rd International Conference on Intelligent Communication Technologies and Virtual Mobile Networks, ICICV 2021, 2021, pp. 512–515, 9388385.

[8] Kim.C.M, Park .H.M, Kim.T, Choi.Y.K, and Lee.S.Y, "FPGA implementation of ICA algorithm for blind signal separation and adaptive noise canceling," IEEE Trans. Neural Network., vol. 14, no. 5, pp.1038–1046, Sep. 2003.

[9] Jolliffe.I.T, Principal Component Analysis. Springer-Verlag, NY, 2002.

[10] Hyv¨arinen.A. One-unit learning rules for independent component analysis: A statistical analysis. Advances in Neural Information Processing Systems, 9:480–486, 1997.

[11] Hyv̈arinen.A Survey on independent component analysis. Neural Computing Surveys, 2:94–128, 1999.
[12] Celik,A. Stanacevic,M. and Cauwenberghs.G,"Mixed-signal realtime adaptive blind source separation," in Proc. IEEE Int. Symp. Circuits Syst., May 2004, vol. 5, pp. V-760–V-763.
[13] Douglas. S.C,Blind source separation and independent component analysis: A crossroads of tools and ideas. In 4th International Symposium on Independent component analysis and blind signal separation (ICA2003), pages 1–10, Nara, Japan, 2003.
[14] R. Kabilan, V. Chandran, J. Yogapriya, Alagar Karthick, Priyesh P. Gandhi, V. Mohanavel, Robbi Rahim, S. Manoharan, "Short-Term Power Prediction of Building Integrated Photovoltaic (BIPV) System Based on Machine Learning Algorithms", International Journal of Photoenergy, vol. 2021, Article ID 5582418, 11 pages, 2021.

Chapter 2

High Performance Fiber-Wireless Uplink for CDMA 5G Networks Communication

R. Ravi[1], R. Kabilan[2], and S. Shargunam[3]

[1]Professor, Department of CSE, Francis Xavier Engineering College, Tirunelveli, India
[2]Associate Professor, Department of ECE, Francis Xavier Engineering College, Tirunelveli, India
[3]Full Time Ph.D Scholar, Dept of ECE, Francis Xavier Engineering College, Tirunelveli, India
Email: fxhodcse@gmail.com; rkabilan13@gmail.com; shargunamguna@gmail.com

Abstract

The demand for bandwidth in mobile communication has been increasing exponentially day by day as the number of users has increased dramatically over the last five years. As an outcome, upcoming wireless communication systems should meet more stringent criteria in order to satisfy customers. Existing wireless systems could only transmit data at a few megabits per second. On the other hand, mm Wave and optical fiber technology have always had the possibility to supply data capacity on the order of Mbps as well as Tbps, respectively. As an outcome, the specifications of such a broadband wireless system could be met by integrating optical fibre and millimeter wave wireless systems. Due to substantial changes in the radio signal, the uplink is greatly hampered. This research establishes the best estimate and subsequent equalization algorithm again for the Fi-Wi CDMA uplink. The equalization is carried out using an innovative Hammerstein category decision feedback equalizer that takes advantage of the properties of PN sequences. This study considers multipath dispersion, nonlinear distortion, noise, and multiuser interference. The correlation characteristics of white-noise similar PN sequences allow for decoupling of both the wireless as well as optical channel parts. In addition, we present a novel approach for mitigating MAI. Simulations of the BER reveal that this technique leaves only a modest amount of residual MAI. Furthermore, power allocation is used to improve CDMA 5G network performance by boosting the signal to noise ratio.

2.1 Introduction

Communication is defined as the conveyance of information from a sender to a recipient across a medium. Communication allows us to be aware of what is going on around us. It allows us to share our knowledge with others while also allowing us to benefit from the thoughts and ideas of others. Communication occurs through a variety of routes and channels, as well as the use of a medium. Digital communication is a type of communication in which data or ideas are encoded digitally as discrete signals and sent to recipients via the internet [6]. In today's world, digital communication is one of the most widely used ways of communication. For the most part, businesses rely on this method for all of their business communications.

The increased efficiency in noisy environments is one of the most important benefits of digital transmission networks for video, data, as well as voice communications over analog counterparts. Inter symbol interference (ISI) is a common by product of digital data transfer. Equalizers play a role in reducing the effects of ISI, which is essential for just a steady digital transmission system [19].

The amplitude as well as phase dispersion of the channel cause the transmitted signals to stop interfering with one another, necessitating the use of equalisers. The assumption that such a channel function generator is known has an impact on transmitter and receiver design [7]. With most digital communications implementations, however, the channel transfer feature is not well acknowledged enough yet to implement filters just at the sender and receiver to minimize channel impact.

The 2G mobile network appeared in the 1990s. In the worldwide market, two systems contended for supremacy: the GSM standard produced in Europe and the CDMA standard developed in the United States. These differed from earlier generations in that they used digital transmission instead of analogue, as well as quick out-of-band network communication. In 2G introduced SMS, or text messaging, as a new mode of communication [17].

Not just stationary users, but also mobile users, benefit from 4G technology's extremely fast wireless internet connectivity. In terms of speed and quality, this technology is projected to overcome the shortcomings of 3G technology. Support for global mobility, as well as integrated wifi and tailored services. Wi-Max and Wi-Fi were used as independent wireless technologies in 3G, but 4G Technologies is planned to combine the two. When 4G technology is fully implemented, the internet will be flooded with HD content, and downloading or high-quality streaming will no longer be an issue. To make consistent use of smart phones, 4G can be effectively coupled with cellular technologies. Because 4G is a multipurpose and versatile technology, it can make use of nearly all packet switching technologies.

Fifth-generation wireless (5G) is the most recent update in mobile communication. Data sent speed will be multi-gigabit speeds, it can also extend up to 20 gigabits per second. To meet the growing reliance on mobile and internet-enabled devices, 5G services will applicable in service for more than several years. As, 5G introduce new applications and users in new domain [10]. Wireless networks have been divided into small sectors and transmission will be happen in radio wave. The foundation stone for 5G is 4G and it is called Long-Term Evolution (LTE) technology. Unlike 4G, which can carry signal for long distance, 5G signal can be transmitted to a long distance with the help of small sectors positioned on building tops[16]. Low frequency bands of spectrum have been used in older version of wireless networks. Poorer-frequency spectrum travels farther but at a slower and lower capacity than MM wave [5].

2.2 Proposed Method

The growth of bandwidth through mobile communication is increasing dramatically day by day as the number of users must have increased dramatically over the last five years. As a result, upcoming wireless systems should indeed meet stricter requirements in order to support a wide range of broadband wireless services. Emerging wireless systems could only transmit data at several megabits per second. mm Wave as well as optical fiber technology, on the other hand, have always had the potential to provide data capacities on the order of Mbps and Tbps. As a result, the needs of such a broadband wireless system could be met by combining optical fibre as well as millimeter wave wireless systems. We propose a modified millimeter wave wireless system based on the OFDM method and a feeder network of optical fiber. Simulated results show that when the SNR rises, the signal becomes less influenced by noise, resulting in fewer decoding errors.

The Internet's exponential growth and the success of 5G systems together with WLANs have had a profound impact on our perception of communication. Transmission speeds such at 100 Mbps are now beginning to reach homes.

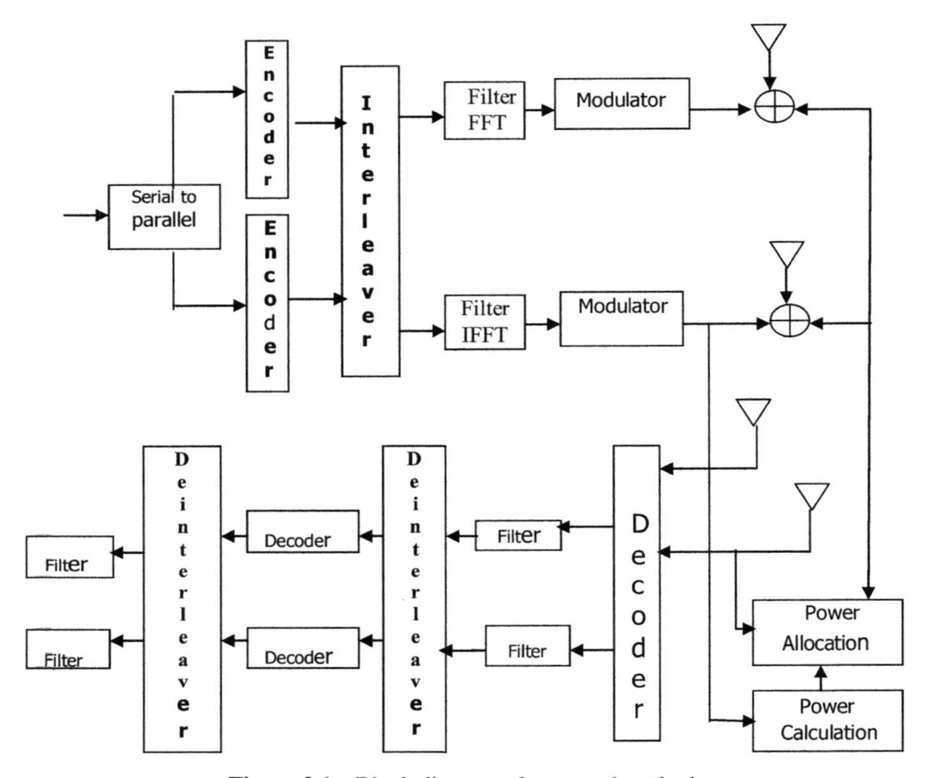

Figure 2.1 Block diagram of proposed method.

QAM will work with a NxN data stream, however the encoder output will be Nx1. The bit interleave is used to transform the encoder output to NxN format. The interleaver's output is fed into the matching filter block, which filters out any noise. The filter's output is modulated and broadcast through the appropriate antennas. During transmission, the noise will be introduced to the channel. The deinterleaver block receives the outputs from the filters. The associated decoder block receives the resultant signal from the deinterleaver block. Another deinterleaver block is used once again. More sounds will be added to the signal during reception; to fix this, we use two deinterleaver blocks, the output of which is provided to the matching filter block. For power allocation, we use the Lagrange theorem and the water filling algorithm. Only the water filling algorithm is used in the channel, resulting in a very low noise ratio at the receiver. We are employing power allocation to boost the SNR, and we will achieve an SNR of more than 20db.

Figure 2.1 illustrate about transmitter and receiver for OFDMA, whereas Figure 2.2 demonstrate proposed system. The general processing

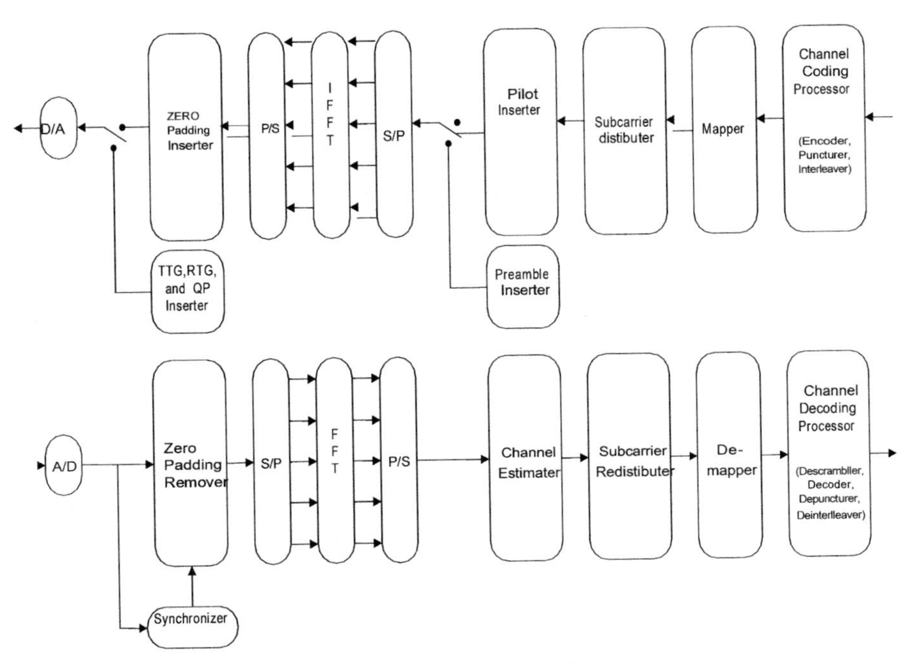

Figure 2.2 Detailed proposed system.

of the WRAN baseband signal is specified in this sub clause. The MAC layer provides the binary data for transmission to the PHY layer. A channel coding processor, which contains a data scrambler, encoder, and bit interleaver, receives this data. For QAM modulation, the interleaved data is transferred to data constellations. The data constellations are assigned to the sub channels by the subcarrier allocator. The preamble is inserted in the first 1 or 2 OFDM symbols of each frame to enable the synchronization, channel estimation, and tracking processes, and the pilot subcarriers are sent at fixed places in the frequency domain within each OFDM symbol.

2.2.1 OFDM

OFDM is specially made for wideband digital communication, It is been used in audio broadcasting, digital television, wireless networks, and 4G 5G mobile communications DSL broadband internet access. An OFDM symbol contains the data, pilot, and null subcarriers in the frequency domain. In essence, the OFDM receiver performs the transmitter's activities in reverse order. The receiver must also perform synchronization and channel estimates in addition to data processing.

2.2.2 OFDMA

OFDMA is a multi-user version of the classic OFDM digital modulation method. As shown in the diagram below, OFDMA accomplishes multiple access by allocating segments of subcarriers to specific users. Multiple users can send low-speed data and information using this method.

2.2.3 CDMA

CDMA stands for spread spectrum multiple accesses. With the same transmitted power, a spread spectrum approach spreads the data bandwidth equally. A CDMA runs faster data rate and the data can be transferred faster.

2.2.4 Optical fiber channel

Cladding mode propagation is also expected to be negligible. Furthermore, we believed that fiber has a lower loss rate. Finally, we assumed that the material and wave-guide dispersive effects are minor in comparison with experimental dispersion. Furthermore, this system shows that the input field amplitude is neither modal dependent nor bulk attenuated as it travels down the fiber, is a real constant.

2.2.5 The disadvantages of the existing system

- In the receiver, tight synchronization between users is essential.
- For synchronizations, pilot signals are employed.
- OFDM is higher difficult than CDMA.
- Advanced coordination among neighboring base stations with dynamic channel allocation. Dynamic routing has a number of advantages over static routing, the most important of which are scalability and adaptability.
- Routers learn about the network topology by talking with other routers using a dynamic router protocol. To the other routers on the network, each router broadcasts its presence and the routes it has available.

2.3 Results and Discussion

2.3.1 Inference 1

A small amount of time is required for each stage of cancellation. As a result, a trade-off must always be decided upon between the number of canceled users and the number of delays that can be tolerated. As a result, a trade-off must be made between power-order precision and acceptable processing

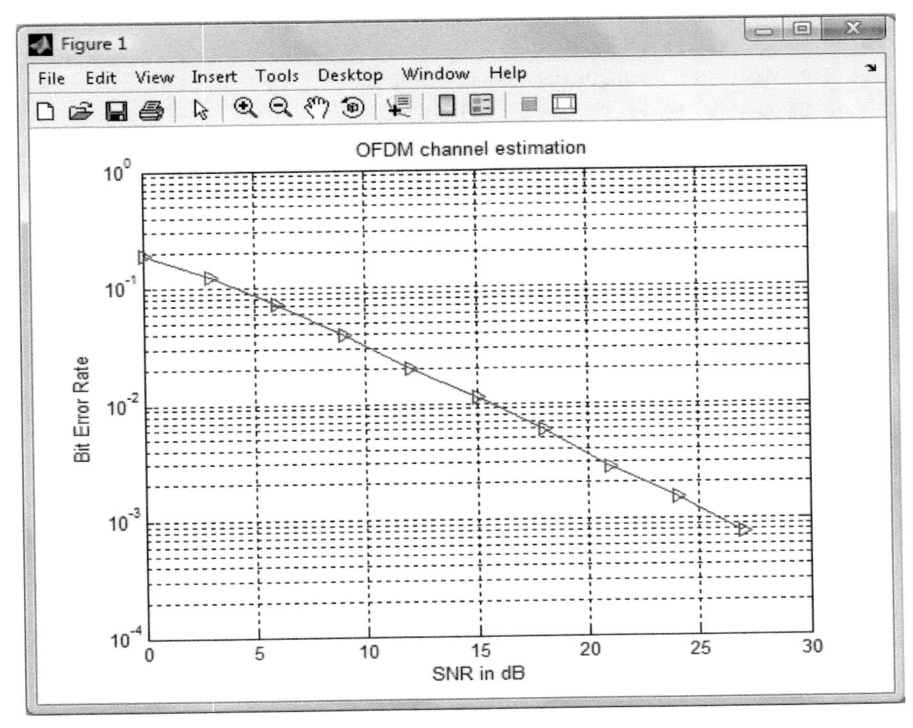

Figure 2.3 OFDM channel estimation.

complexity. If indeed the bit assumption is incorrect, the bit's interfering effect mostly on the SNR ratio is quadrupled in power. When compared to performance, this same implementation of HDFE resulted in a reduction in bit error rates. Multi-user identification can also be used to enhance receiver performance in channels with large delay spreads.

2.3.2 Inference 2

As shown in figure 2.4, the iterative methods of the CIR assumption translate into a more accurate assessment of the polynomial. This is primarily due to the iterative algorithm's ability to remove classes of non-erroneous peaks and more accurately assess the actual CIR peaks. Because of this interconnected effect, even just a slight increase in the CIR estimate has a significant impact on the polynomial estimate.

2.3.3 Inference 3

On MIMO-OFDM channels evaluated in various indoor environments, the BER and PER performance of MIMO-OFDM to bit unsettled coded

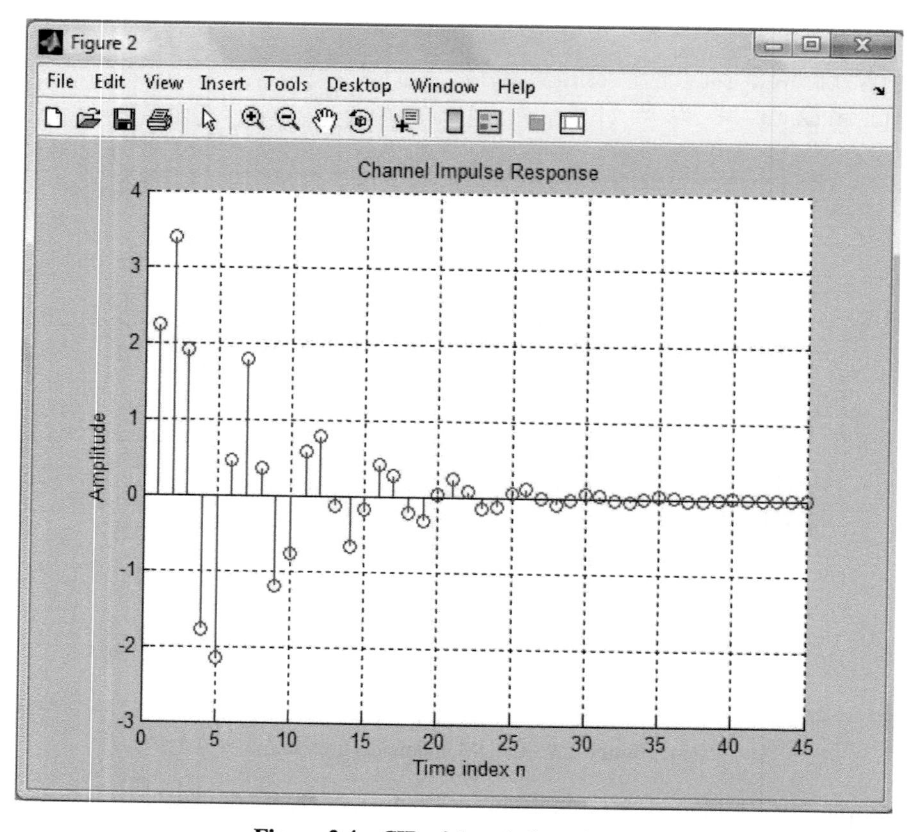

Figure 2.4 CIR of the wireless channel.

modulation MIMO-OFDM-BICM devices are evaluated. The achievement of MIMO-OFDM-BICM in such a LoS environment could really differ wildly depending on the presence of scattering, whereas this is more evenly spread in an NLoS environment, which also has SNR per receiver.

2.3.4 Inference 4

Here for fiber-wireless uplink after estimating the channel we are reaching more than 25dB SNR. As the SNR increases the error reduces up to or more than 25dB.

2.3.5 Inference 5

Here we used 2X2 antenna that is each antenna can transmit up to two receivers. This is the advantage of using QAM modulation. Since we are using

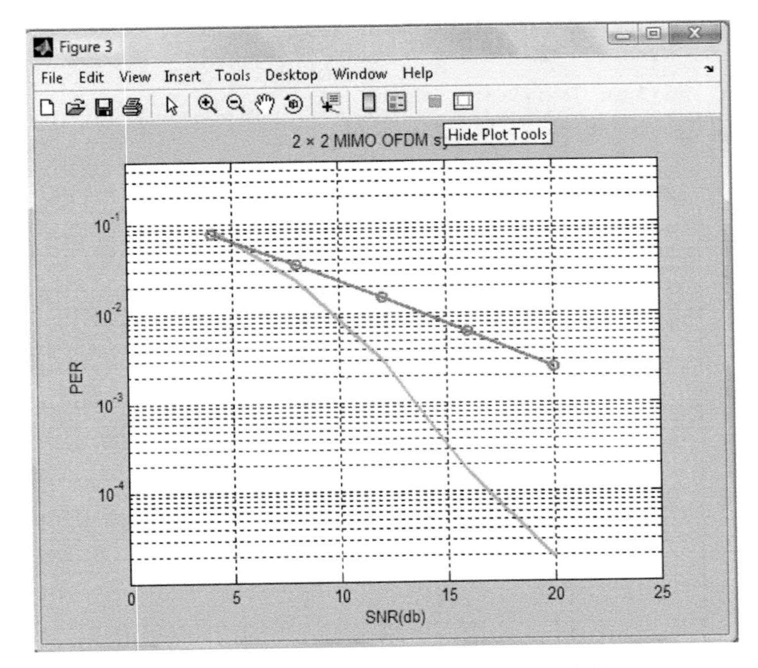

Figure 2.5 SNR Vs PER of MIMO OFDM for 2X2 users.

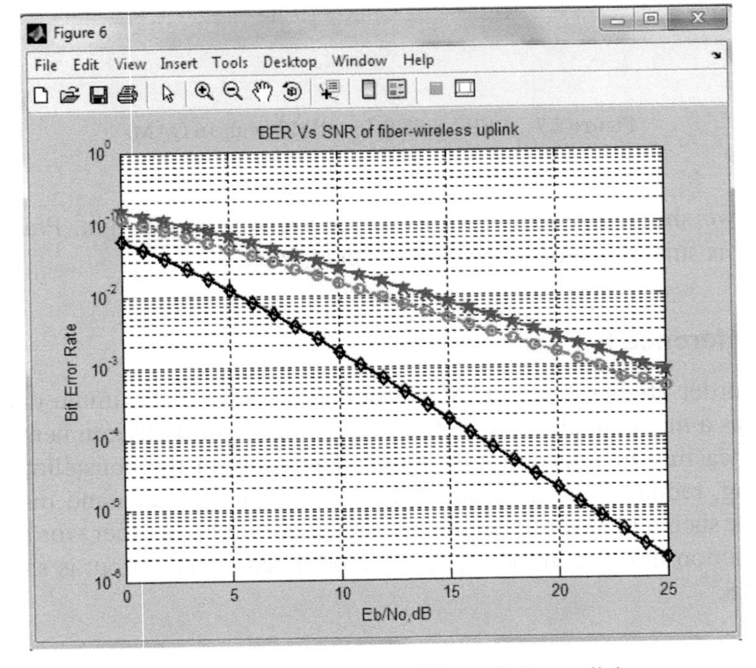

Figure 2.6 BER Vs SNR of fiber-wireless uplink.

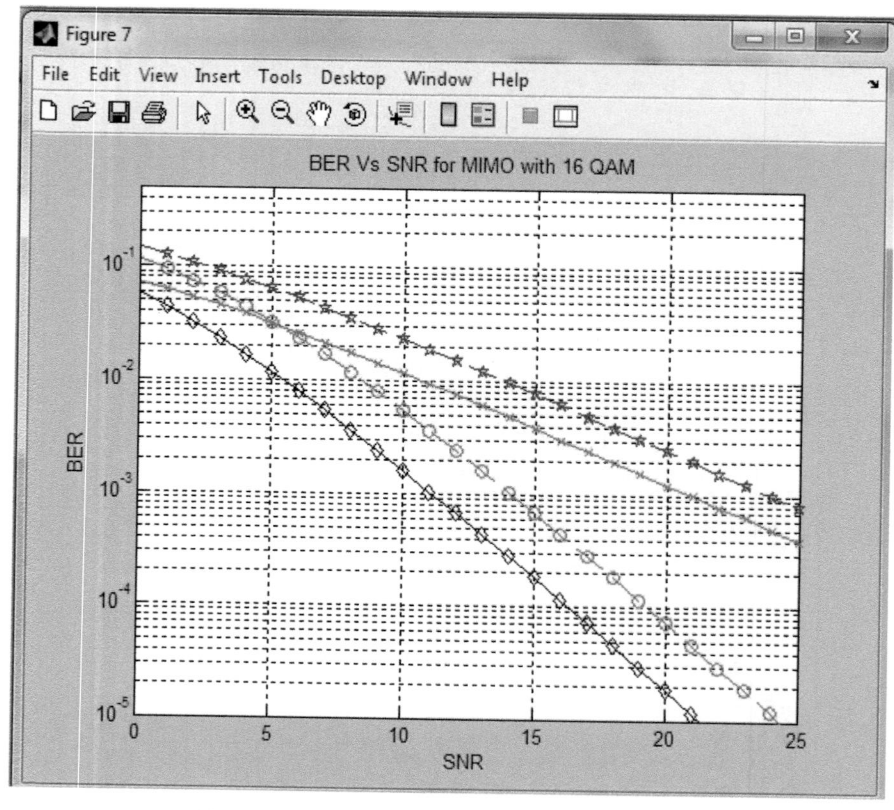

Figure 2.7 BER Vs SNR for MIMO with 16 QAM.

MIMO we should use QAM for SISO QPSK will be all right. Practically 16QAM is suitable compared to BPSK and QPSK.

2.3.6 Inference 6

Higher order Higher-order QAM constellations have a maximum data rate as well as a mode in adversarial RF or microwave QAM environments, such as broadcasting or telecommunications. The spots in the constellation are spreading, reducing the detachment between adjacent states and making it tough for such a receiver to decode the signal correctly. In other words, there is a reduction in noise immunity.QAM modulated signal output is shown in figure 2.8.

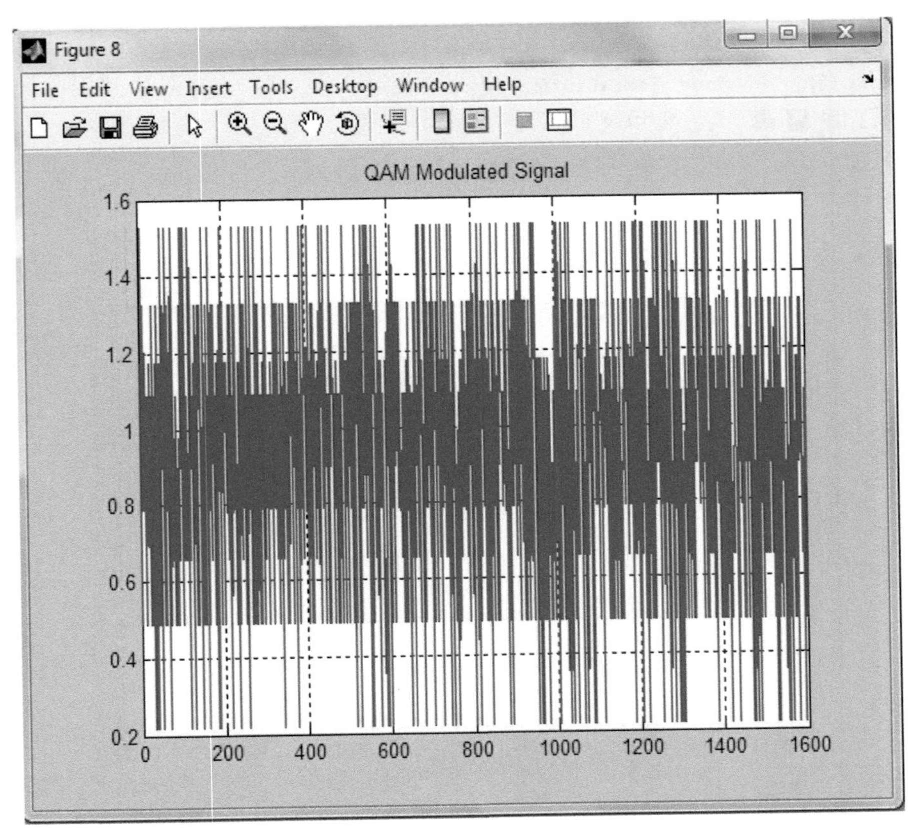

Figure 2.8 QAM modulated signal.

2.3.7 Inference 7

The result shows in figure 2.9 illustrate the coverage area of the signal transmitted. No. of iterations implies distance in Kilometer. Here we can infer that the signal is covering more than 20Km. For better performance we need to place one repeater receiving antenna after every 20Km distance.

2.4 Conclusion

An efficient method for identifying and equalizing the uplink in such a multi-user CDMA Fi-Wi network has been proposed in this paper. Estimation was accomplished through the use of PN sequence similarity features, and equalization was accomplished through the use of a single equaliser with

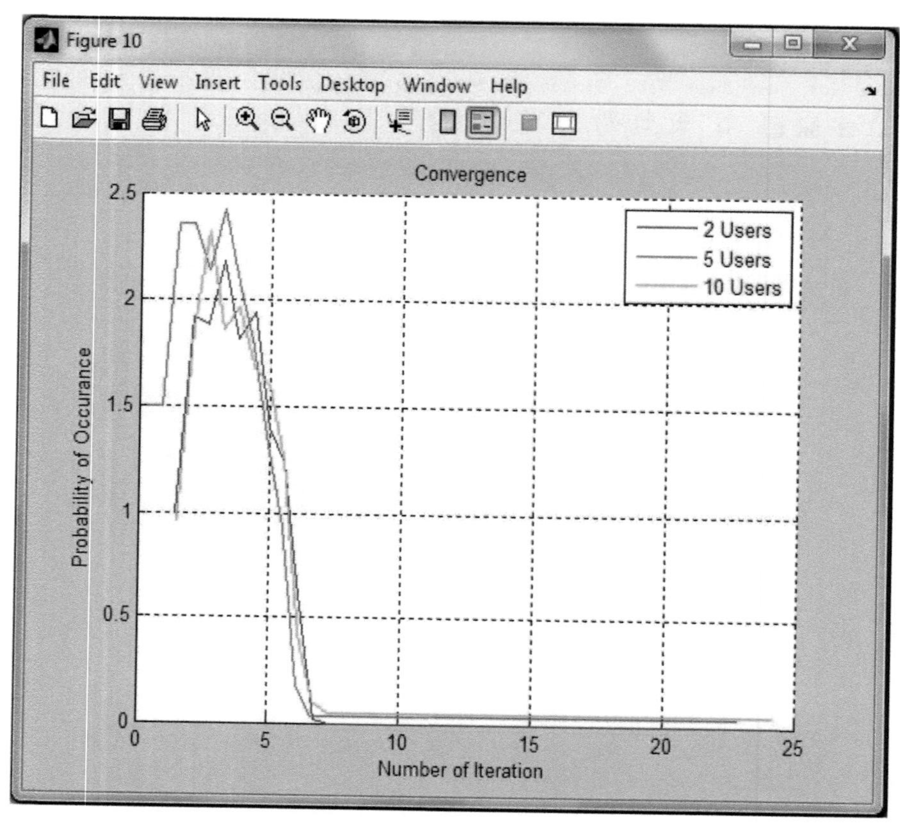

Figure 2.9 Convergence graph.

different linear and non-linear modules. The algorithm also mitigates MAI. This method is suitable for a 4G uplink because it functions in such an asynchronous CDMA environment. The ability to estimate and equalize the linear and nonlinear elements separately is a key advantage of this method. The non-linear estimation does not have to be replicated frequently because of the nonlinearity of both the ROF link and the data. It only changes if the temperature of the optical transmitter or other physical conditions change dramatically. The linear wireless channel, needs both equalizers as well as frequent estimation. Furthermore, each user's wireless channel is unique, whereas nonlinear ROF link have been shared among all the users. Quality of CIR estimate was shown to be dependent on the CIR's properties. The proposed approach achieves an SNR of over 25db. The severity of multipath situations affects BER performance in general. A good BER for data transmission is

106, which our technique can accomplish with an SNR of around 25 dB. Ultimately, the above technique is suited to any asynchronous CDMA link to multipath dispersion as well as moderate nonlinearity. This nonlinearity could be due to the base station's low noise amplifier. This may seem unusual because most study focuses on the transmitter amplifier's nonlinearity. Even through this, the handset's transmitter amplifier just controls a single user signal inside the CDMA uplink. The receiver amplifier must be linear but also handle a huge multiuser signal.

References

[1] F. Adachi, D. Garg, S. Takaoka, and K. Takeda,(2005) "Broadband CDMA techniques," IEEE Wireless Commun., vol. 12, no. 2, pp. 8–18.

[2] Chengjin Zhang, Member, IEEE, and Robert R.Bitmead, Fellow, IEEE, (2008) "MIMO Equalization With State-Space Channel Models",IEEE TRANSACTIONS ON SIGNAL PROCESSING,VOL.56, NO.10.

[3] Djebbar A.B, Abed-Meraim K, and Djebbar.A,(2009), "Blind and Semi Blind Equalization of Downlink MC-CDMA System Exploiting Guard Interval Redundancy and Excess Codes", IEEE TRANSACTIONS ON COMMUNICATIONS,VOL.57. NO.1

[4] Don Torrieri, Senior Member, IEEE, Amitav Mukherjee, Student Member, IEEE, and Hyuck M. Kwon, Senior Member, IEEE,(2010),"Coded DS-CDMA Systems with Iterative Channel Estimation and no Pilot Symbols "IEEE TRANSACTIONS ON WIRELESS COMMUNICATIONS, VOL.9, NO.6.

[5] Farooq Khan, "Performance of orthogonal uplink multiple access for beyond 3G/4G systems," in Proc. IEEE Veh. Technol. Conf., Piscataway, NJ, USA, 2006, pp. 699–704.

[6] Jong-Seob Baek and Jong-Soo Seo,Member,IEEE,(2008),"Efficient Design of Block Adaptive Equalization and Diversity Combining for Space-Time Block- Coded Single-Carrier Systems", IEEE TRANSACTIONS ON WIRELESS COMMUNICATIONS, VOL.7, NO.7.

[7] Justin Coon, Student Member, IEEE, Simon Armour, Member, IEEE, Mark Beach, Associate Member, IEEE, and Joe McGeehan,(2005), "Adaptive Frequency-Domain Equalization for Single-Carrier Multiple-Input Multiple- Output Wireless Transmissions", IEEE TRANSACTIONS ON SIGNAL PROCESSING,VOL.53, NO.8.

[8] Mahesh.K.Varanasi, Member, IEEE, And Behnaam Aazhang, Member, IEEE, (1990), "Multistage Detection in Asynchronous

Code-Division Multiple-Access Communications"IEEE TRANSAC COMMUNICATIONS,VOL.38. NO.4.

[9] Martin Wrulich, Student Member, IEEE, Christian Mehlfuhrer, Member,IEEE, and Markus Rupp, Senior Member, IEEE,(2010)," Managing the Interference Structure of MIMO HSDPA: A Multi-User Interference Aware MMSE Receiver with Moderate Complexity, "IEEE TRANSACTIONS ON WIRELESS COMMUNICATIONS, VOL.9, NO.4.

[10] MicheleMorelli,Member,IEEE,LucaSanguinetti,StudentMember,IEEE, and Umberto Mengali, Life Fellow, IEEE, (2005), "Channel Estimation for Adaptive Frequency-Domain Equalization", IEEE TRANSACTIONS ON WIRELESS COMMUNICATIONS, VOL.4, NO.5.

[11] Muck.M, Courville. M.D, and Duhamel.P,(2006), "A Pseudo-random postfix OFDM modulator semi-blind channel estimation and equalization", IEEE TRANSACTIONS ON SIGNAL PROCESSING,VOL.52, NO.16.

[12] Paul G. Flikkema, Department of Electrical Engineering University of South Florida Tampa, Florida USA,(1997), Decision Feedback Multipath Cancellation for Coherent Direct Sequence Spread Spectrum Wireless", IEEE TRANSACTIONS ON WIRELESS TECHNOLOGY, VOL.19, NO.24.

[13] Pinter S.Z and Fernando X.N, (2007)," Concatenated fibre-wireless channel identification in a multiuser CDMA environment, IEEE TRANSACTIONS ON COMMUNICATIONS,VOL.38. NO.4.

[14] Ricardo Merched, Senior Member,(2010), "Fast Algorithms in Slow And Higher Doppler Mobile Environments", IEEE TRANSACTIONS ON WIRELESS COMMUNICATIONS, VOL.49, NO.4.

[15] Stephen Z.Pinter, Student Member, IEEE, and Xavier N. Fernando, Senior Member, IEEE,(2010), "Estimation and Equalization of Fi-Wi Uplink for Multiuser CDMA 4G Networks, "IEEE TRANSACTIONS ON COMMUNICATIONS,VOL.58. NO.4.

[16] Stephen Z.Pinter and Xavier N. Fernando,IEEE,(2005), "Estimation Of Radio- Over Fiber Uplink In a Multiuser CDMA Environment Using PN Spreading", CCECE/CCGEI, Saskatoon,VOL.34. No.5.

[17] Tingting Liu, Student Member, IEEE, Chenyang Yang, Senior Member, IEEE, and Lie-Liang Yang, Senior Member, IEEE,(2010), "Joint Transmitter-Receiver Frequency-Domain Equalization in Generalized Multicarrier Code-Division Multiplexing Systems", IEEE TRANSACTIONS ON VEHICULAR TECHNOLOGY,VOL.59. NO.8.

[18] Tsung-Hui Chang, Member, IEEE, Wei-Cheng Chiang, Y.W Peter Hong, Member, IEEE, and Chong-Yung Chi, Senior Member, IEEE,(2010), "Training Sequence Design for Discriminatory Channel Estimation in Wireless MIMO Systems" IEEE TRANSACTIONS ON SIGNAL PROCESSING,VOL.58, NO.12.

[19] Veeraruna Kavitha and Vinod Sharma, (2006), "Tracking Performance of an LMS-Linear Equalizer for Fading Channels", Forty-Fourth Annual Allerton Conference House, UIUC, Illinois, USA Sept 27–29.

[20] Wenkun Wen, Minghua Xia, and Yik-Chung Wu,(2010), "Low Complexity Pre- Equalization Algorithms for Zero-Padded Block Transmission" IEEE TRANSACTIONS ON WIRELESS COMMUNICATIONS, VOL.9, NO.8.

[21] Xavier.N.Fernando, Member, IEEE, and Abu B. Sesay, Senior Member, IEEE, (2002), "Adaptive Asymmetric Linearization of Radio Over Fiber Links for Wireless Access "IEEE TRANSACTIONS ON VEHICULAR TECHNOLOGY,VOL.51. NO.6.

Chapter 3

Improving the Performance of Cooperative Transmission Protocol Using Bidirectional Relays and Multi User Detection

R. Kabilan[1], R. Ravi[2], and S. Shargunam[3]

[1]Associate Professor, Department of ECE, Francis Xavier Engineering College, Tirunelveli, India
[2]Professor, Department of CSE, Francis Xavier Engineering College, Tirunelveli, India
[3]Full Time Ph.D Scholar, Dept of ECE, Francis Xavier Engineering College, Tirunelveli, India
Email: rkabilan13@gmail.com; fxhodcse@gmail.com; shargunamguna@gmail.com

Abstract

Cooperative transmission is a new communication strategy that takes advantage of the broadcast aspect of wireless channels to increase transmission by allowing network nodes to cooperate. Its potential for further application in wireless networks is limited due to its low spectral efficiency and the requirement for orthogonal channels. Using MUD, a cooperative transmission protocol with good spectrum efficiency, diversity gain, and coding gain was developed in this study. The benefit of MUD cooperative transmission methods is that improving one user's link can help other users. By completely exploiting the link between the relay and the destination, the coding and high diversity order can be attained. In this case, one of the nodes will operate as a relay, facilitating collaboration among the network's various nodes. MUD will determine which user will send the data and send it to the appropriate destination. Each node in our project will operate as a relay for any other node. Data loss and packet retransmission will be reduced as a result of this. The network's performance will improve as a result of this. We utilize a dynamic base station to ensure that nodes outside the network broadcast packets to the network's destination.

3.1 Introduction

Wireless networks give a rising number of laptop and PDA users greater freedom and mobility, eliminating the need for wires to stay connected to their office and the Internet. The devices that give wireless service to these clients, ironically, require a lot of cables to connect to private networks and the Internet [8]. The wireless mesh network is presented in this white paper as a feasible alternative to all those wires [5,6]. In particular, over the last ten years, the mobile radio communications industry has grown by orders of magnitude, largely due to advances in digital and RF circuit fabrication, new large-scale circuit integration, and other miniaturization technologies that make portable radio equipment smaller, cheaper, and more reliable [2-4]. Digital switching techniques have made it possible to create low-cost, easy-to-use radio communication networks on a broad scale. During the next decade, these tendencies will accelerate even more.

3.2 Components of Communication System

Figure 3.1 describes the components in the communication systems, which are explained below.

- A message is sent from the source, which could be a human voice, a television image, or data.
- For efficient transmission, the transmitter changes the baseband signal. A pre-emphasize, a sampler, a quantizer, a coder, and a modulator are just a few of the subsystems that make up a transmitter.
- The transmitter output is conveyed through a channel, which could be a wire, a coaxial cable, an optical fiber, or a radio link, among other things.
- By removing the signal alterations made at the transmitter and the channel, the receiver reprocessed the signal received from the channel.
- The output transducer receives the receiver output and converts the electrical signal back to its original form. Transmitters and receivers are carefully built to avoid distortion and noise.

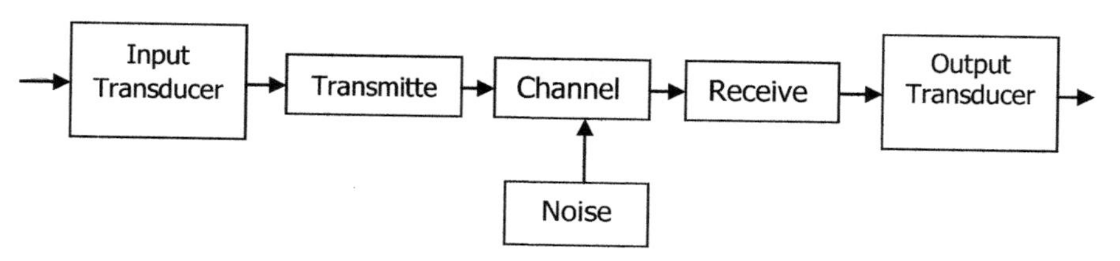

Figure 3.1 Components of communication system.

A brief history of the evolution of mobile communications around the world is useful to grasp the immense influence that cellular radio and PCS will have on all of us over the next many decades [10]. It's also beneficial for a beginner in the cellular radio business to comprehend the enormous influence that government regulatory agencies and service rivals have on the development of new wireless systems, services, and technologies [1]. While the purpose of this work is not to discuss the technological and political implications of cellular radio and personal communications, governments regulate radio spectrum usage, not service providers, equipment makers, entrepreneurs, or researchers [7]. A government's progressive involvement in technology development is critical if it intends to keep its own country competitive in an ever-changing field [9].

3.3 Proposed System

This system includes a cooperative transmission technique as well as multi-user detection. Each node will serve as a relay for the next node. To compute the shortest distance to the destination, we use the AOMDV algorithm. The path with the shortest distance is taken into account. Relay nodes are used for long-distance communication. In communication, this type of bidirectional communication is used. This system's key benefit is that several users can send and receive data at the same time. In nodes with numerous packets, there are no packet collisions. With this system, we can achieve high efficiency and little packet loss. The advantage of bidirectional communication is also utilized in this system. We employ a dynamic base station in the system, which allows us to acquire a huge coverage area. The data can be sent by multiple people at the same time.

The nodes numbered 0 to 8 in the following block diagram serve as transceiving and relay nodes for long-distance communication. The ninth node will serve as a dynamic BS.

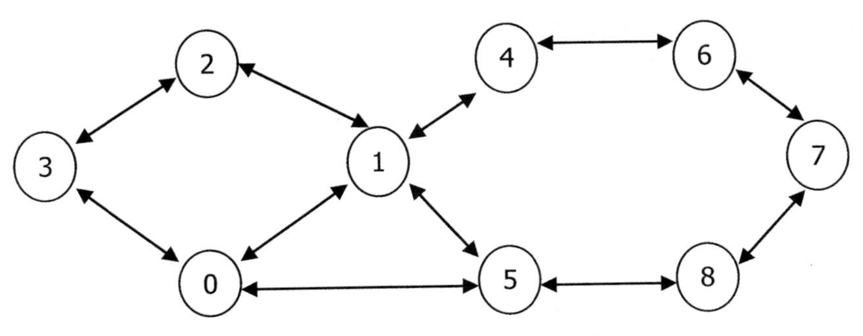

Figure 3.2 Block diagram for proposed system.

3.4 System Design and Development

3.4.1 Input design

Input Design is a component of overall system design, and it necessitates meticulous attention. If the data entering the system is wrong, the processing and output will exaggerate the faults. There are three types of inputs in the system.

External: These are the system's primary inputs;
Internal: These are user communications with the system.
Interactive: These are inputs entered during a conversation with the computer.

3.4.2 Feasibility analysis

Given infinite resources and time, all tasks are feasible. Before moving on to the software development processes, the system analyst must determine whether the proposed system is realistic for the firm and establish client requirements. The fundamental goal of a feasibility study is to see if the problem is worth addressing.

3.4.2.1 Operational feasibility

An operational feasibility assessment is required as part of the feasibility analysis. This is due to the fact that, according to software engineering principles, operational feasibility, or usability, should be extremely high. The system was verified to be operating after a thorough investigation.

3.4.2.2 Technical feasibility

The system analyst will examine the proposed system's technical feasibility. The technical feasibility study takes into account the hardware utilized for system creation, data storage, processing, and output.

3.4.2.3 Economical feasibility

Before proceeding any further with the suggested system's development, the system analyst must assess the proposed system's economic feasibility, and the cost of operating the system must be balanced against the cost savings that can be realized by deploying the system.

3.4.2.4 Project modules

In our project, we use NS to analyze the performance of our system. We investigated 0–9 nodes, with 0–8 nodes acting as both sending and receiving nodes. Dynamic BS is defined as the 9th node. To begin, all of the nodes must be initiated. If a node needs to transfer data, it sends the data over a direct link; otherwise, it uses the AOMDV algorithm to determine the shortest distance to the destination and then sends the data to the appropriate destination.

Nodes 0 and 8 were used as the source and destination nodes, respectively. First, node 0 sends data to node 8, and then the data is transported using the AOMDV method to find the shortest path to the target. The size of a packet is determined by the quantity of data it contains. After a given period of time, node 8 will begin transmitting packets as well. Nodes 6 and 2 have received data from both sources at the same time, and we now have an improved bidirectional connection because of the cooperative transmission protocol.

3.5 Output Design

We get both graphics and animator output when we utilize NS. We created a graph for packet lifetime and an output graph in our system. With the use of a lifespan graph, we can learn more about how long it takes a packet to reach its destination and how effectively it gets there without colliding with other packets from different nodes.

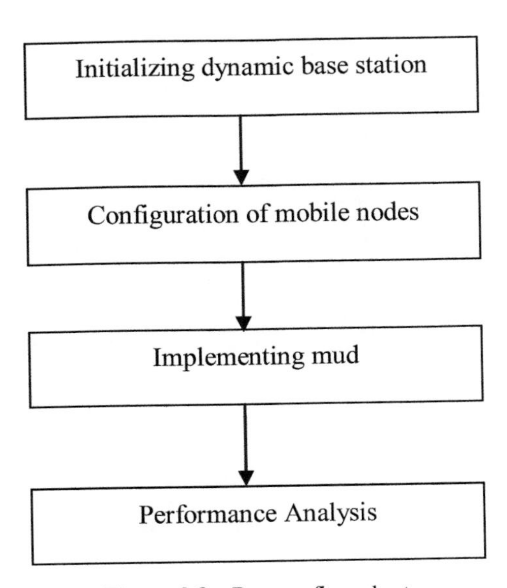

Figure 3.3 Process flow chart.

3.5.1 Animator output

In our system, we'll use a wireless PAN network with 0–9 nodes, with 0–8 nodes acting as transmitting and receiving nodes, and the 9th node acting as a dynamic BS. If one of the nodes wants to deliver a packet to the farthest node, the other node will operate as a relay node. The command window for running the code in Ns is shown in this diagram.

3.5.2 Initialization of nodes

This is the first step in which all the nodes are initialized. We considered nodes 0 and 8 as a source and destination nodes and vice versa for this process takes some amount of time.

3.5.3 Node 1 starts transmitting data

The circles in this figure show the coverage area. After the completion of every transmission the acknowledgement is received.

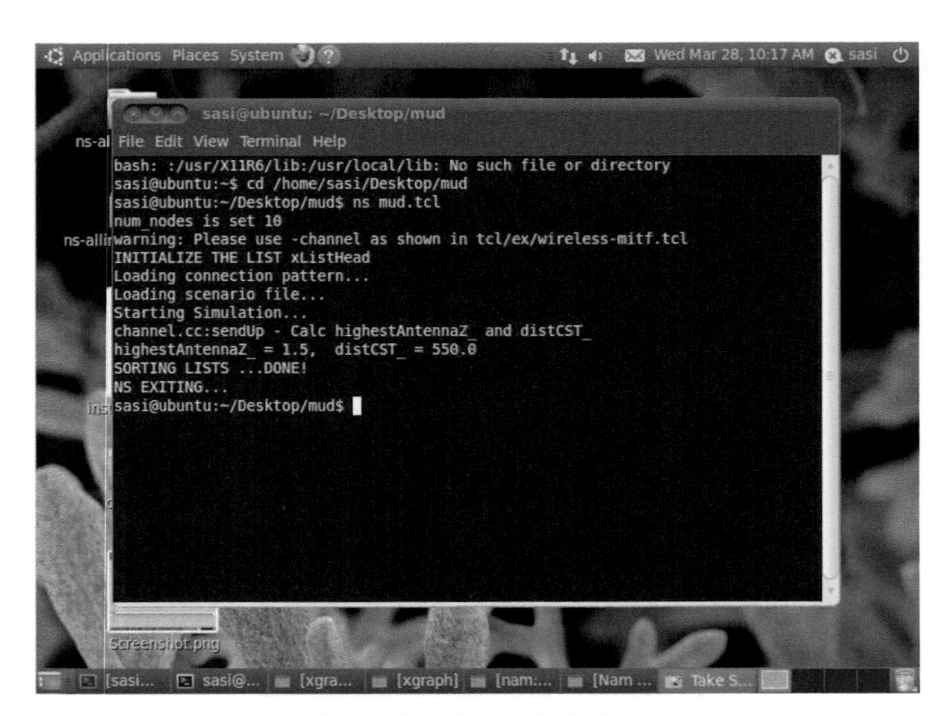

Figure 3.4 Command window.

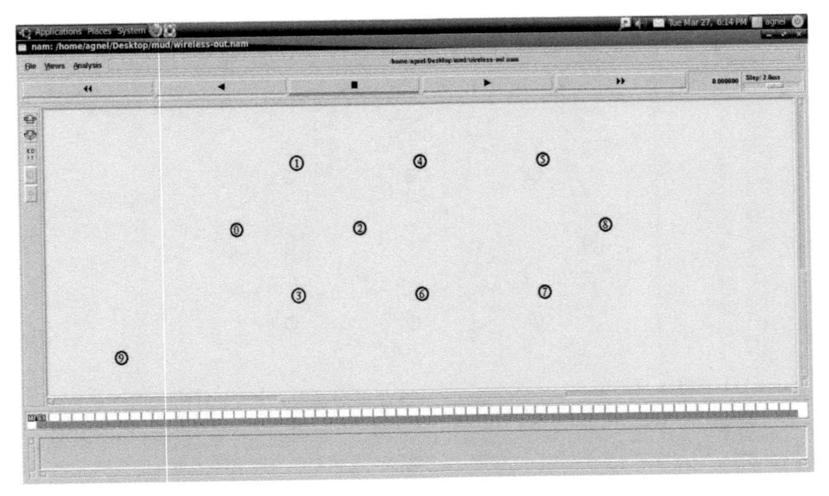

Figure 3.5 Initialization of nodes.

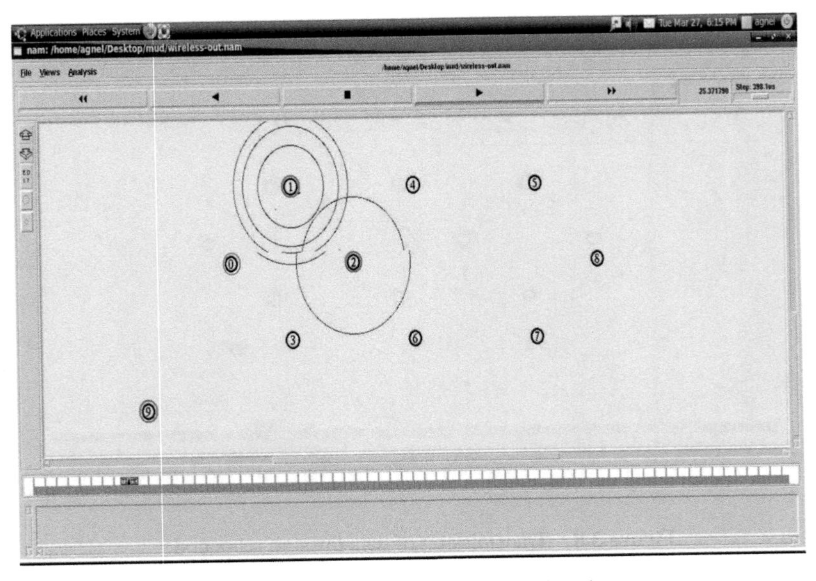

Figure 3.6 Node 1 starts transmitting data.

3.5.4 Finding shortest path

By the use of AOMDV the relay node chose the shortest path to reach the destination. The transmission of data through the relay node is shown in the following screenshots.

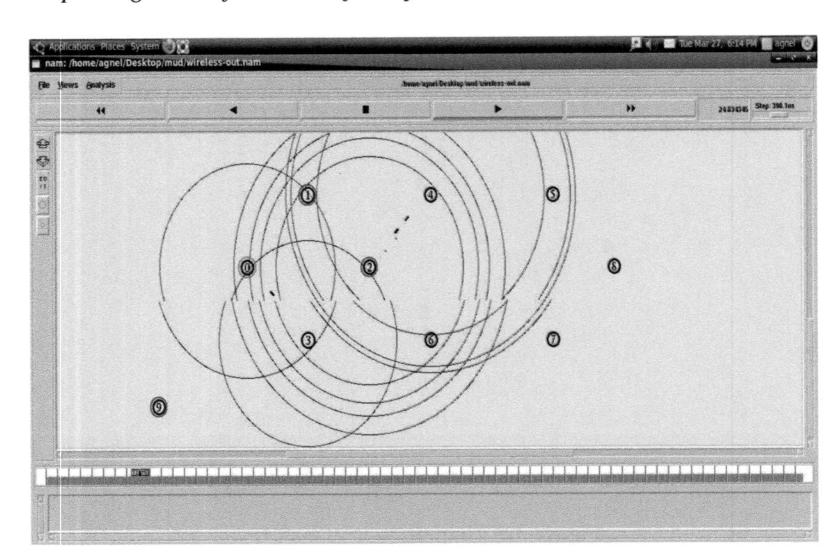

Figure 3.7 Choosing the shortest path.

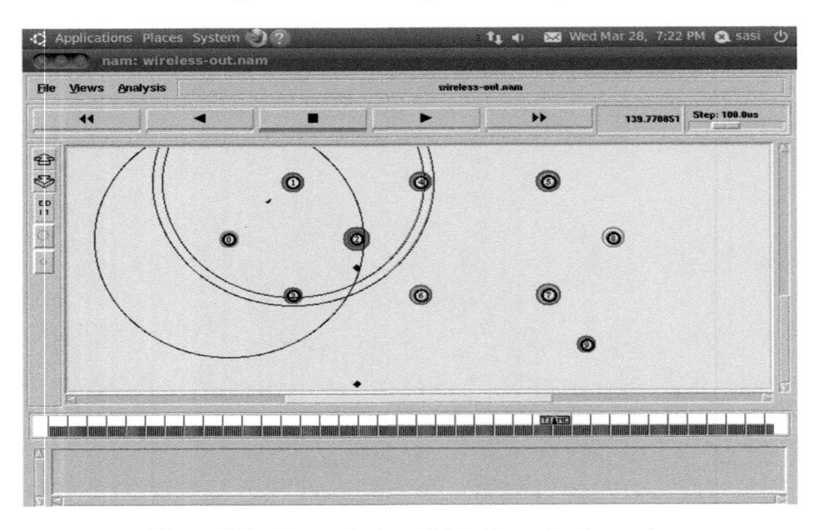

Figure 3.8 Transmission of data through relay node.

3.5.5 Transmission of data through relay node

In this figure node, 1 starts transmitting data to node 8. It sends the data packets to the node which has the shortest path to the destination (i.e.) by using the AOMDV algorithm.

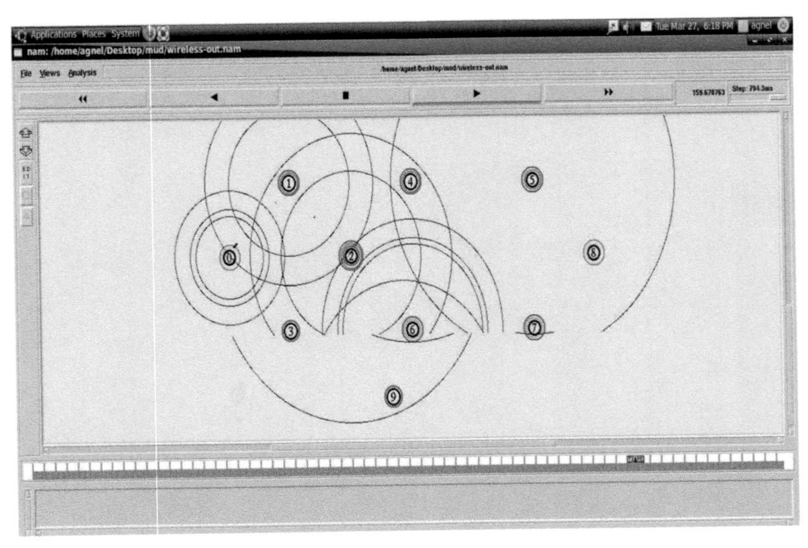

Figure 3.9 Node 8 starts transmitting data.

3.5.6 Node 8 starts transmitting data

In this figure shows that after sometime if node 8 needs to send the data it calculate the shortest distance after the completion of calculating the shortest distance the node 8 itself act as a source and start transmitting data.

3.5.7 Loss of packets

In this figure loss of packets due to collision among packets from both direction.

3.5.8 Transmissions of data from node 7 to node 6

In this figure node, 7 is considered as a shortest distance from node 8 so it sends its data packet to node 7 it relays that to node 6 until the destination is reached.

3.5.9 Transmision of data from node 2 to node 4

In this figure, the node1 start transmitting the nex set of data and the process is similar to as described in figure 5.5

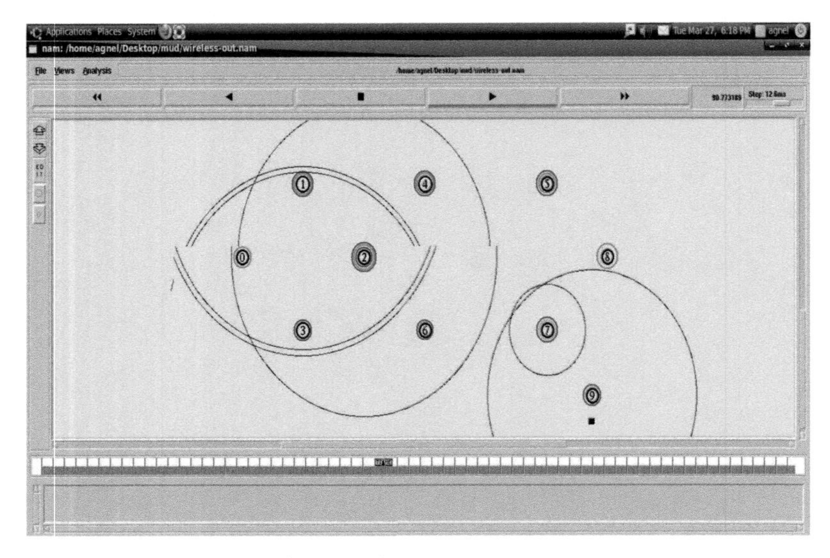

Figure 3.10 Loss of packets.

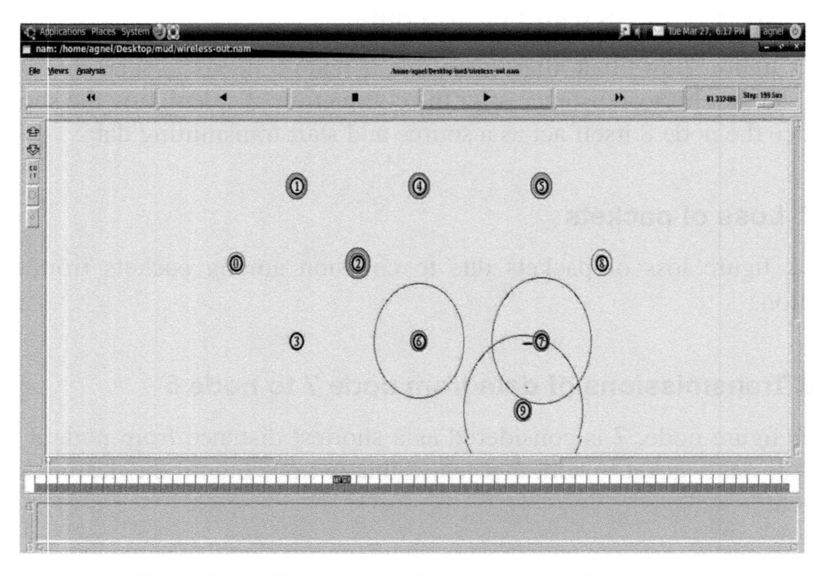

Figure 3.11 Transimision of data from node 7 to node 6.

3.5.10 Transmission of data bidirectionally

In this figure shows the how effectively the packets are sent from both the direction considering nodes 1and 8 as the source.

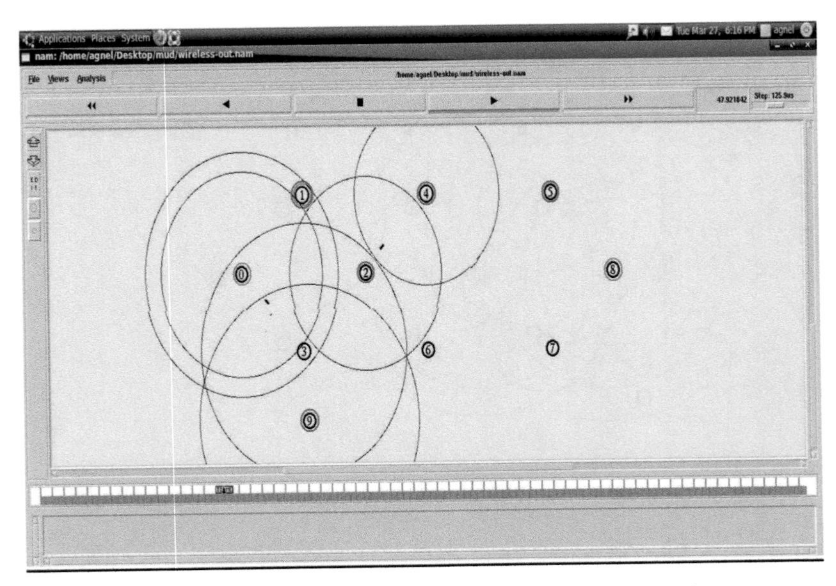

Figure 3.12 Transmision of data from node 2 to node 4.

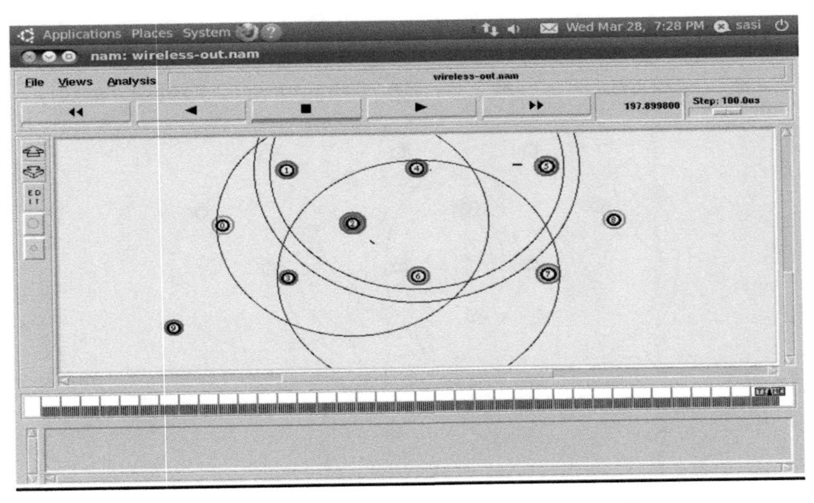

Figure 3.13 Transmission of data bidirectionally.

3.5.11 Completion of transmission from node 8 to 0

This figure shows the completion of transmission of packets which considered node 8 as source and node 0 as a destination.

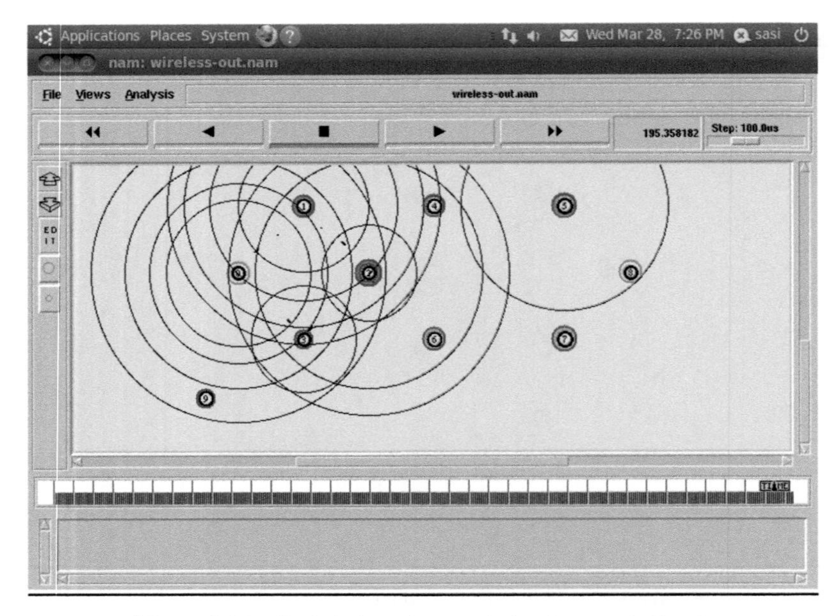

Figure 3.14 Completion of transmission from node 8 to 0.

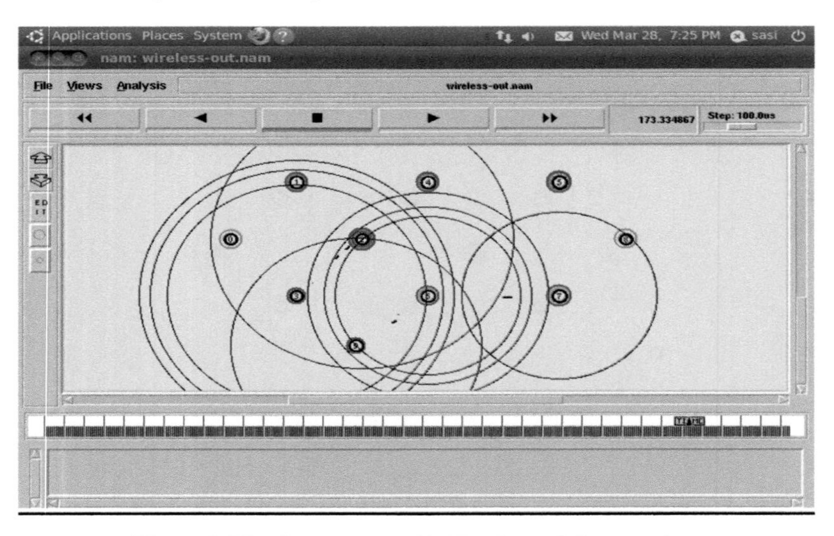

Figure 3.15 Coverage provided by dynamic base station.

3.5.12 Coverage provided by dynamic base station

This figure shows for providing coverage area the dynamic BS moves from its position towards node 8 which has a lack of coverage area.

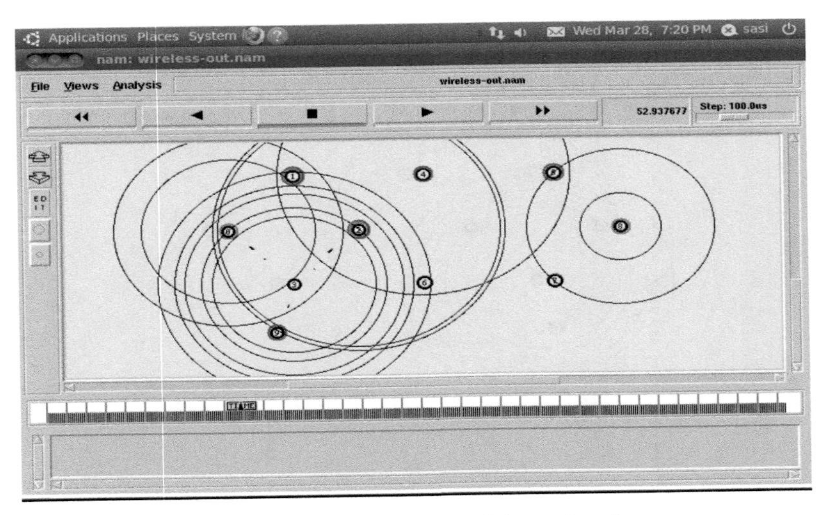

Figure 3.16 Retransmission of dropped packets.

3.5.13 Retransmission of dropped packets

This figure shows that the packets which are loosed during transmission due to collision are retransmitted this process is stopped until the acknowledgment for the loosed data is received.

3.5.14 Reception of acknowledgement

To provide coverage the dynamic BS moves toward node 8 after completion of data transfer it moves towards its initial position.

3.5.15 X graph for lifetime

The output for the packet lifetime and the time at which the packets begin transmitting after the node initialization time are obtained from this graph. After an average of roughly 9 seconds, the nodes are initialized.

The graph was created using the information gleaned from the graph below. The time at which the nodes begin transmitting data packets is indicated by the blue line. The graph's lower spikes represent packet loss in intermediate relay nodes.

3.5.16 X graph for output

From this graph, we obtained the throughput that got from our predefined factors such as the number of nodes etc.,

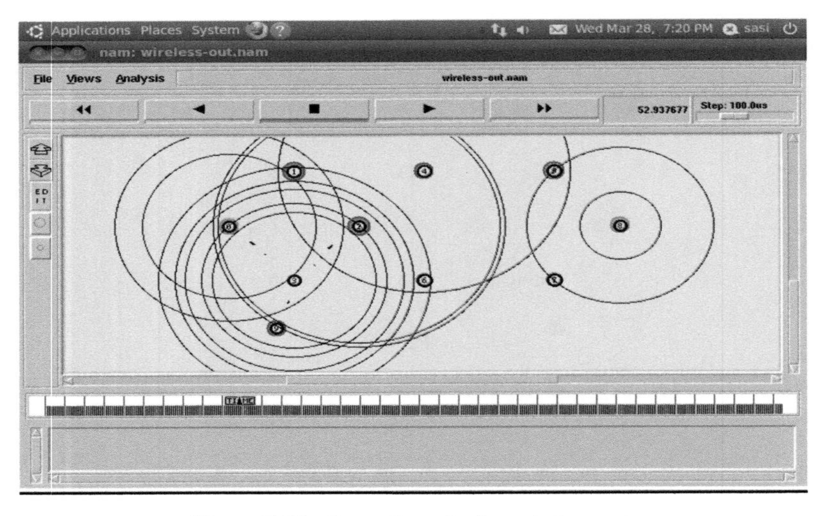

Figure 3.17 Reception of acknowledgement.

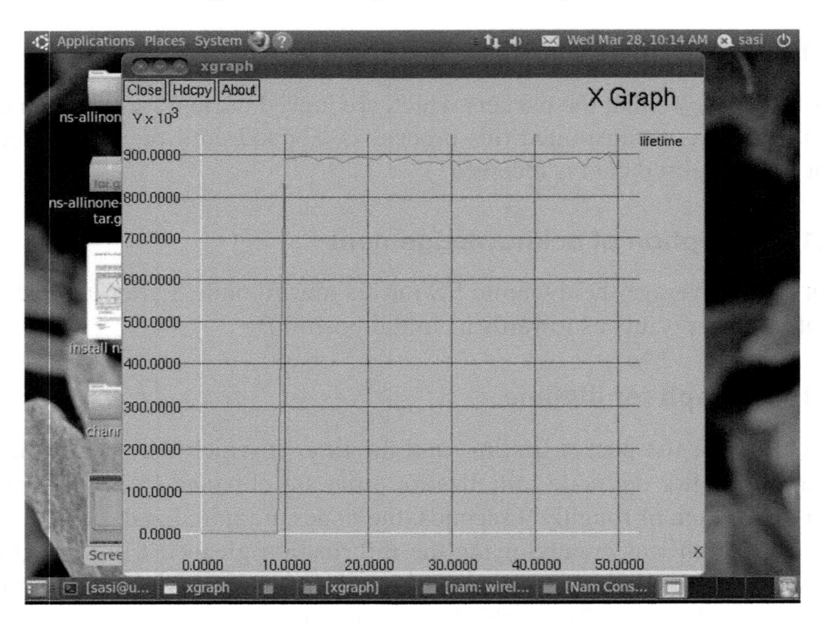

Figure 3.18(a) Graph for lifetime.

We discovered a maximum throughput of roughly 12*103 from the X graph. This will be used to determine how well data is transported to the destination. The value obtained from the method is higher than the throughput obtained from the prior way.

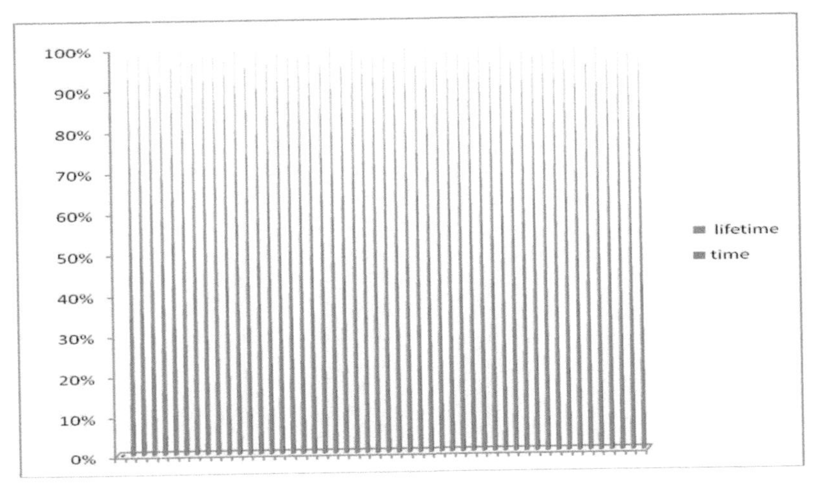

Figure 3.18(b) Lifetime vs node ready.

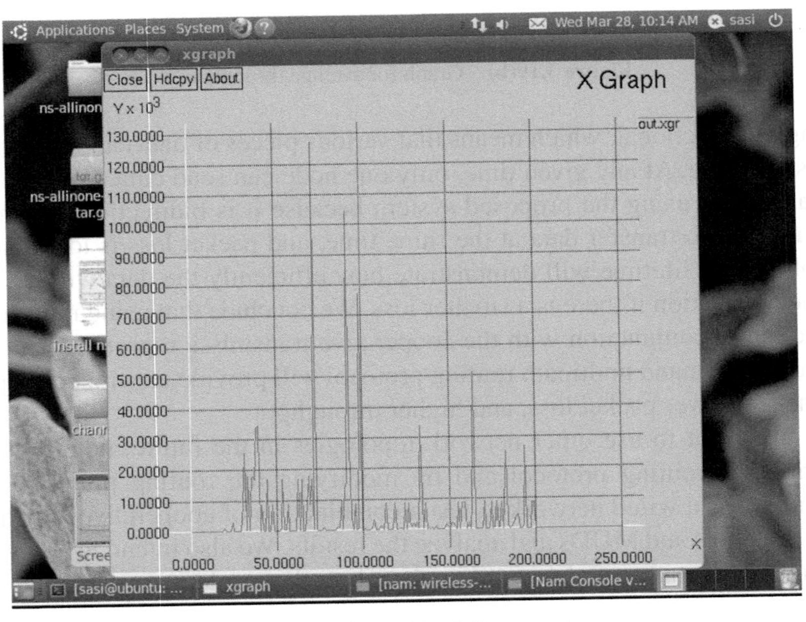

Figure 3.19(a) Graph for output.

3.6 Conclusion

Based on the above sources, we analyze the system performance in each paper. Each existing system we examine has some disadvantages when compared to the suggested system. The existing approach has the problem of

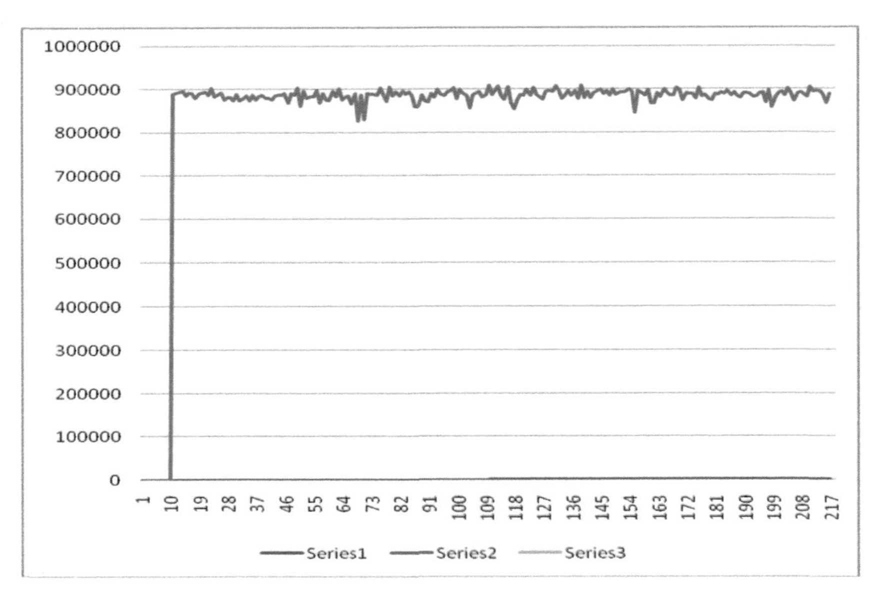

Figure 3.19(b) Graph for average throughput.

being unidirectional, which means that various pieces of data can't be sent at the same time. At any given time, only one node can send data. We have the advantage of using the proposed system because it is bidirectional, numerous nodes can transfer data at the same time, and packet loss is low. The X graph for the lifetime will demonstrate how efficiently the data will be sent to the destination if there is a smaller loss. We conclude that using multi-user detection in conjunction with the cooperative transmission protocol and the ad hoc on-demand multipath routing protocol will provide us with better performance, lower packet loss, and higher throughput.

We want to use other network topologies in the future, both without altering the routing protocol and by modifying the routing protocol. We intend to use a wired network to test a combination of cooperative transmission protocols and MUDs and analyze the results. We also intend to apply the technology to any sort of network and analyze the results.

References

[1] Zhu Han, Xin Zhang, and H. Vincent Poor (MAY 2009) "High Performance Cooperative Transmission Protocol Detection and Network Coding", IEEE Trans., Wireless Communication., Vol 8, No. 5, Pg No.1–5.

[2] A. Sendonaris, E. Erkip, and B. Aazhang, (Nov. 2003) "User cooperation diversity—part I: system description," IEEE Trans. Communication., vol. 51, no. 11, pp.1927–1938.

[3] A. Sendonaris, E. Erkip, and B. Aazhang (Nov. 2003) "User cooperation diversity—part II: system description," IEEE Trans. Communication., vol. 51, no. 11, pp.1927–1938.

[4] Andrew Sendonaris, Member, IEEE, Elza Erkip, Member, IEEE, and Behnaam Aazhang, Fellow, IEEE, (November 2003) "User Cooperation Diversity—Part I: System Description" IEEE transactions on communications, vol. 51, no. 11.

[5] J. N. Laneman, D. N. C. Tse, and G. W. Wornell (Dec. 2004) "Cooperative diversity in wireless networks: efficient protocols and outage behavior," IEEE Trans. Inform. Theory, vol. 50, no. 12, pp. 3062–3080.

[6] Z. Yang, J. Liu, and A. Host-Madsen, (Nov. 2005) "Cooperative routing and power allocation in ad-hoc networks," in Proc. IEEE Global Telecommunications Conference, Dallas, TX.

[7] Ahmed K. Sadek, Zhu Han, and K. J. Ray Liu, Department of Electrical and Computer Engineering, and Institute for Systems Research , University of Maryland, College Park, MD 20742, USA. "A Distributed Relay Assignment Algorithm for Cooperative Communications in Wireless Networks", IEEE ICC 2006 proceedings ,pg no1592–1597

[8] Zhiguo Ding, Member, IEEE, Kin K. Leung, Fellow, IEEE, Dennis L. Goeckel, Senior Member, IEEE, and Don Towsle, Fellow, IEEE, ,.(August 2010)"A Relay Assisted Cooperative Transmission Protocol for Wireless Multiple Access Systems", IEEE transactions on communications, vol. 58, no. 8.

[9] Le Quang Vinh TRAN, Olivier BERDER, Olivier SENTIEYS IRISA/ INRIA, University of Rennes 1 ,6 rue de Kerampont BP 80518 - 22305 Lannion Cedex, France ,"Non-Regenerative Full Distributed Space-Time Codes in Cooperative Relaying Networks" IEEE 2011,pg no 1529–1533.

[10] Hong Zhou, Faculty of Engineering, Osaka Institute of Technology, Osaka. Japan." Combination of Bi-directional and Cooperative Relaying for Wireless Networks", 2010, 17th International Conference, pg no 331–338.

Chapter 4

Joint Relay-source Escalation for SINR Maximization in Multi Relay Networks and Multi Antenna

R. Ravi[1], R. Kabilan[2], S. Shargunam[3], and R. Mallika Pandeeswari[4]

[1]Professor, Department of CSE, Francis Xavier Engineering College, Tirunelveli, India
[2]Associate Professor, Department of ECE, Francis Xavier Engineering College, Tirunelveli, India
[3]Full Time Ph.D Scholar, Dept of ECE, Francis Xavier Engineering College, Tirunelveli, India
[4]Full Time Ph.D Scholar, Dept of ECE, Francis Xavier Engineering College, Tirunelveli, India
Email: fxhodcse@gmail.com; rkabilan13@gmail.com; shargunamguna@gmail.com; mallikapandeeswari123@gmail.com

Abstract

This study presents a very effective strategy for increasing SINR, QoS and lowering source and relay transmit power. A single relay was employed between the sender and the receiver in the prior procedure. It adds to the optimization difficulty. Multi-relay networks have been employed between the source and the prefixed receiver in this suggested system to aid in the search for the best solution. In this situation, the half-duplex protocol is used. Precoders for the source and relay are found in the source and prefixed receivers, respectively. The source precoder is affected by the Second Order Cone Program issue, whereas the relay precoder is affected by the Semi Definite Relaxation problem. These issues may be solved using the Interior Point Algorithm, which comprises the rank-one matrix method and the randomization technique. This may be done with the help of MATLAB. In the future, this might be expanded to include the full-duplex protocol for more efficient mobile communication.

4.1 Main Text

1. This method's purpose is to merge relay precoders and sources on a receiver, which seems to be tough both practically and theoretically. It has two optimized methods for both relay precoders and the source for a prefixed receiver. SINR will be lowered in the receiver output, proportional to the relay transmit source and power. Under severe relay power restrictions and source [3,] the SINR will be reduced. It lowers the source–relay power while maintaining the second criterion's QoS requirement. This strategy preserves QoS while lowering the overall amount of electricity optimized [21].

2. Both situations result in non–convex optimization challenges. The number of antennas at each node must be equal to the number of signal strengths, according to the feasibility of both optimization problems. Iterative interchanging approaches are provided to handle the two independent optimization concerns, in which the precoders are calculated differently for each iterative process, i.e., all precoders are fixed save the one that is optimized. The optimum control issue for source precoder data processing may be stated as a SOCP problem, with the interior point technique [2] as an alternate solution.

3. Other electromagnetic wireless devices, such as light, magnetic, and electric fields, as well as sound, are less prevalent techniques for establishing wireless communications [18]. Wireless operations make it possible to provide services that would be difficult or impossible to provide using cables, such as long-range communications. The term refers to telecommunications systems that transmit data without the need of wires and with the use of some type of energy. Data is sent across short and long distances using this technology [19-20].

4. From a few meters to hundreds of kilometers, transmission distances vary. Wireless communication equipment includes cordless phones, mobile phones, GPS systems, wireless computer components, and satellite television [1]. Some of the advantages of wireless communication are as follows: Communication has increased in order to provide clients with more timely information. Working professionals may work and access the Internet from anywhere, at any time, without the need for cables or connections. This also helps to keep the project on time and boosts productivity. Doctors, laborers, and other professionals operating in distant areas may interact with medical centers via wireless communication. People may be alerted to an emergency situation through wireless communication [16-17]. Through wireless transmission, these

alerts may be utilized to give aid and support to the affected communities. Wireless networks are less costly to set up and operate. Because of the emergence of wireless networks, we can now utilize personal gadgets anywhere and at any time[8-14].

5. This has benefited humanity in developing in many parts of life, but it has also brought up a slew of dangers. As a consequence of the wireless network, several security concerns have emerged. Wireless signals that are transmitted in the air may be readily intercepted by hackers. It's vital to keep the wireless network secure so that unauthorized users can't access the data [7,15]. This also increases the risk of data loss. Strong security protocols like WPA and WPA2 must be created to protect wireless communications. The main objectives of this project are to enhance throughput and QoS by increasing SINR and lowering source and relay transmit power [4-6].

4.2 Proposed System

To tackle two non–convex problems, iterative alternating methods are utilized; in this case, we use precoders alternately in each iteration to fix all precoders and save the one that was optimized. The optimum solution to the source precoder data processing optimization issue, which may be shown as a SOCP problem, is the interior point system. Similarly, relay precoder problems will be a semi-definite relaxation issue, therefore ready–to–use methods will be tuned to tackle it.

4.2.1 System model

The relay network contains source, destination, and M relays, R1,..., RN, as shown in Figure 4.1. NS, ND, and NR antennas are installed on the destination and the source nodes. The relays will have the same antenna number for notational convenience. This method can be easily extended with relays and it has more numbers of antennas.

Every broadcast is divided into two time slots by a half-duplex protocol. The signal from the source node will be received in the first time slot. The transfer signal from the relay filter will be received in the second time slot. At destination, the second slot signal will be received. Between the destination and the source, node has no direct connections. On the transmitter side, the transmitted signal will distinguish spatial multiplexing and signal streams. The relays are completely synced, and the source has full CSI. Each relay node will send a training sequence to relays from the source node. Relay and Destination will use error-free feedback channels to convey the estimated channels back to the source.

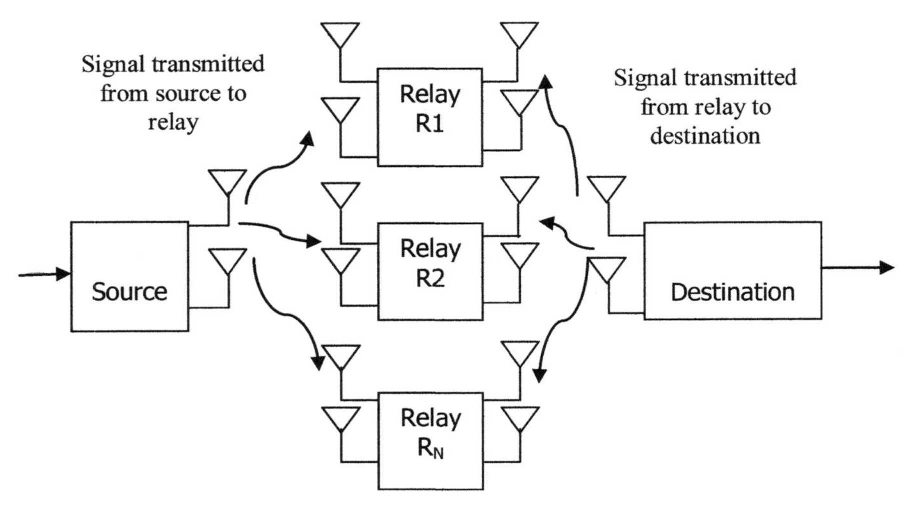

Figure 4.1 Relay network with M relays, source, and destination.

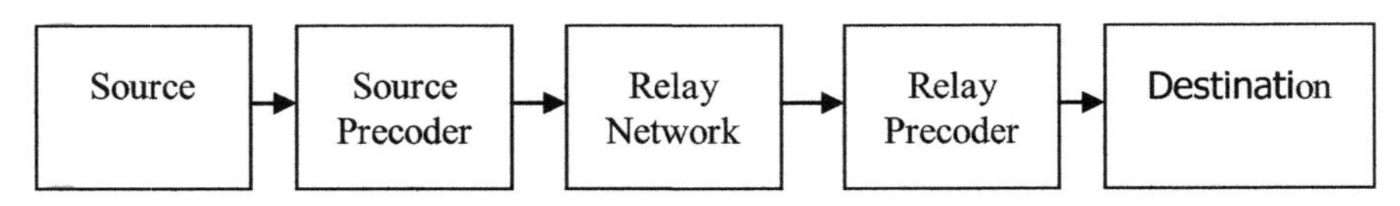

Figure 4.2 Skeletal view of proposed method.

The source node is where the precoders are computed. Figure 4.2 shows the skeletal view of the proposed method. The receiver filter coefficients must be sent between source to destination to the components destination filters, relay as well as source to be jointly optimized. The relay as well as source in the sections using two different methods, source transmit power reduction, source to relay transmit power reduction, and SINR, with QoS constraints.

4.2.2 SINR maximization under relay transmit power and source constraints

The source as well as a relay are often located spatially and use their own power supplies. It is a more realistic scenario than a relay power constraint and source. For all relays, a shared transmit power constraint is considered.

4.2.3 Source-relay transmit power minimization under QoS constraints

QoS is ensured with the minimum transmit power achievable in some applications. The main theme is to reduce the transmit power between the relays and the source. The iterative alternate approach is used to solve it. In specifically, the precoders are computed alternately in each iteration; in each iteration, all precoders are fixed except, but the nonprecoders got optimized.

4.2.4 Computation of relay precoder

The optimization problem is NP-hard because it is a non–convex fractional QCQP. It is proposed to address it by converting it into an SDP known as SDR. A simple bisection search can quickly reveal the ideal. It's worth mentioning that the solution you get isn't always of rank one.

4.2.5 Feasibility of the problem

The feasibility of the optimization challenge is verified before detailing how to optimize the source and relay precoders. It is ensured that SIR, rather than SINR, is used to justify the recommended problems; however, SINR is used for precoder design.

4.3 Advantage

- Increase the spectral efficiency
- Increase Reliability
- It minimizes the power usage
- It maximization of the worst streams

4.4 Application

- Television broadcasting
- Radio broadcasting

4.5 Result and Discussion

4.5.1 Tools used

Physical component resources, often known as hardware, are the most prevalent set of needs given by any Operating System or software program. A Hardware Compatibility List (HCL) is frequently included with a hardware requirements list, particularly in the case of operating systems. A Hardware Compatibility List (HCL) was used to verify compatible and sometimes incompatible hardware devices for a specific Operating System or application.

4.5.2 Simulated results

This figure 4.3 shows that whenever the SNR value increases, the bit error rate gets reduced with each relay.

The suggested technique performs better when compared with separate power transmit constraints.

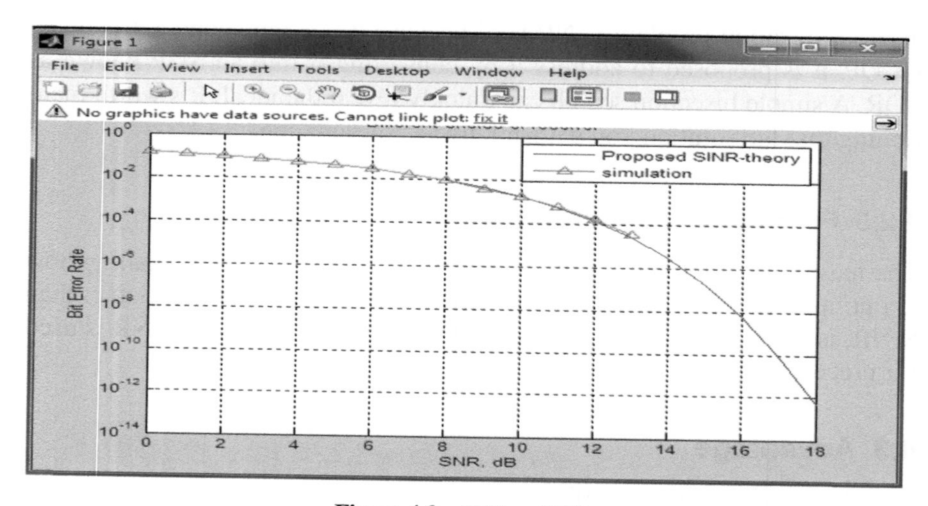

Figure 4.3 SNR vs BER.

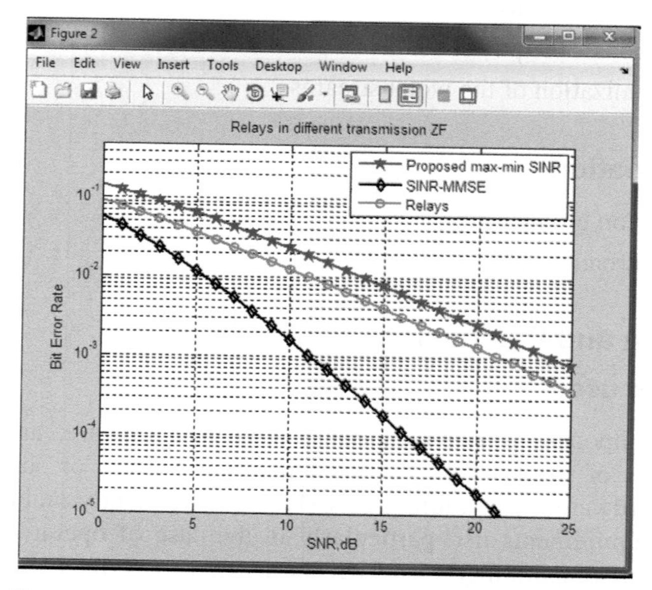

Figure 4.4 Shows that the relation between the SNR and BER between various antenna connected to relays.

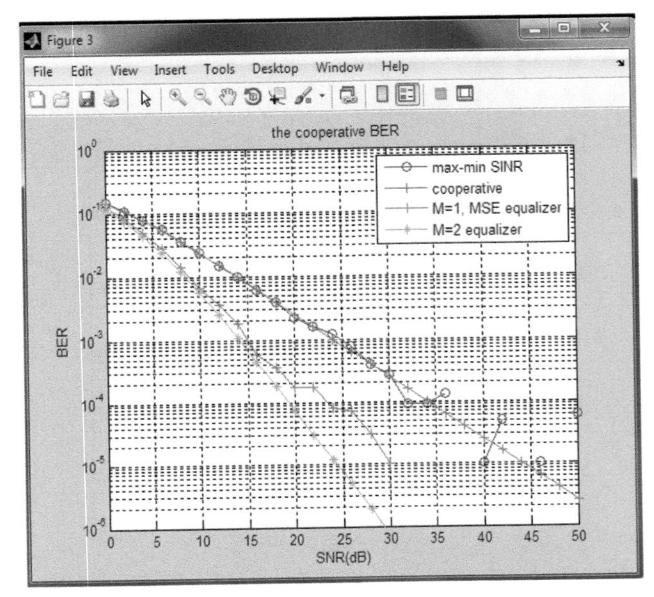

Figure 4.5 Shows that M = 1, L = 2 relay signal level where every node got connected with two antennas and obtain the minimum bit error rate.

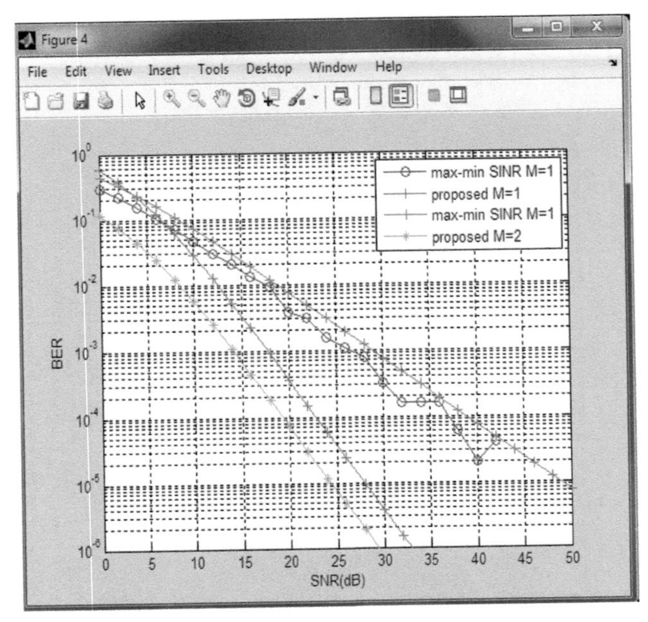

Figure 4.6 Shows that the relationship between SNR and BER when different number of relays are used.

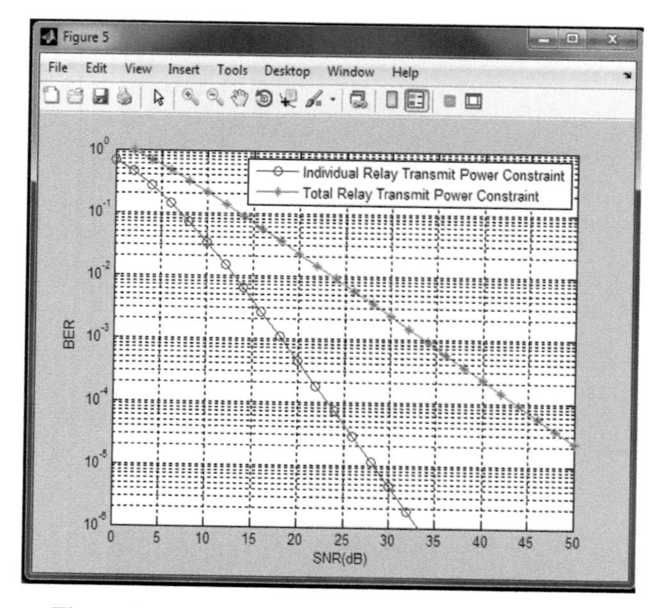

Figure 4.7 Shows that total relay transmit power limits.

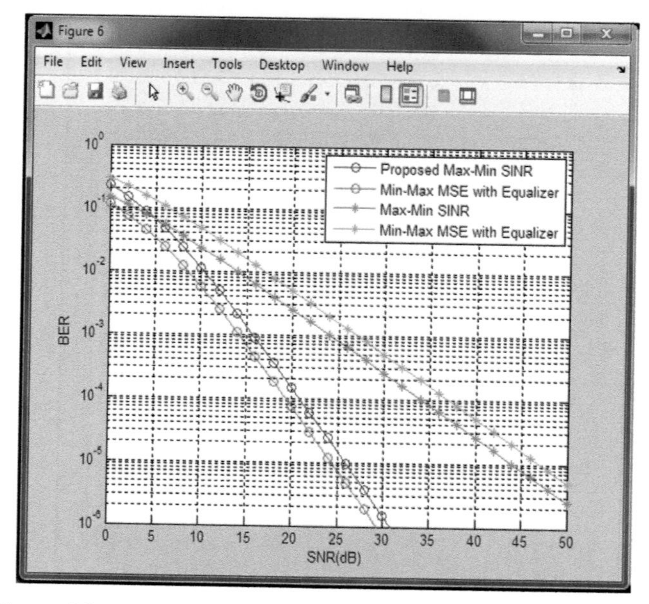

Figure 4.8 Shows that the proposed SINR method Max-Min Level.

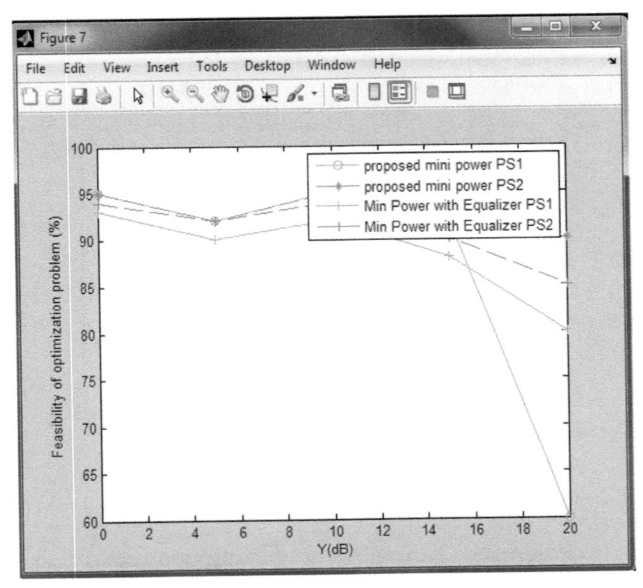

Figure 4.9 It shows how increasing the transmit power of both sources enhances the suitability of both the optimization processes and decreases the bandgap between them.

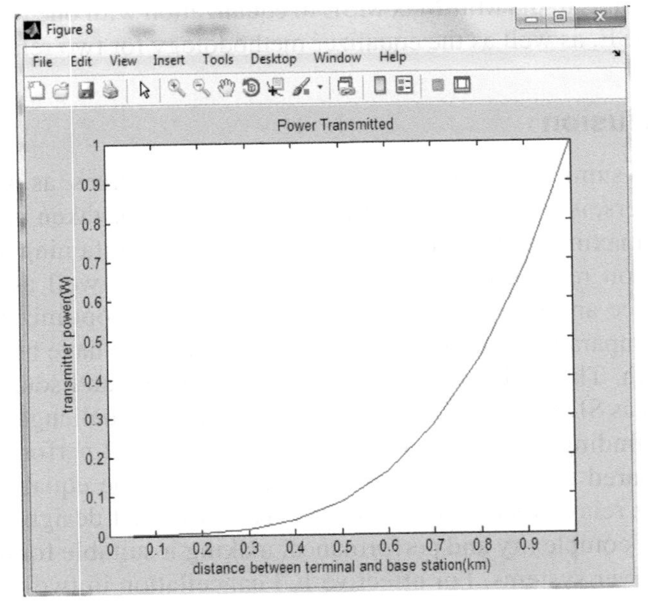

Figure 4.10 Indicates that for long-distance broadcasting, the smallest possible power is necessary.

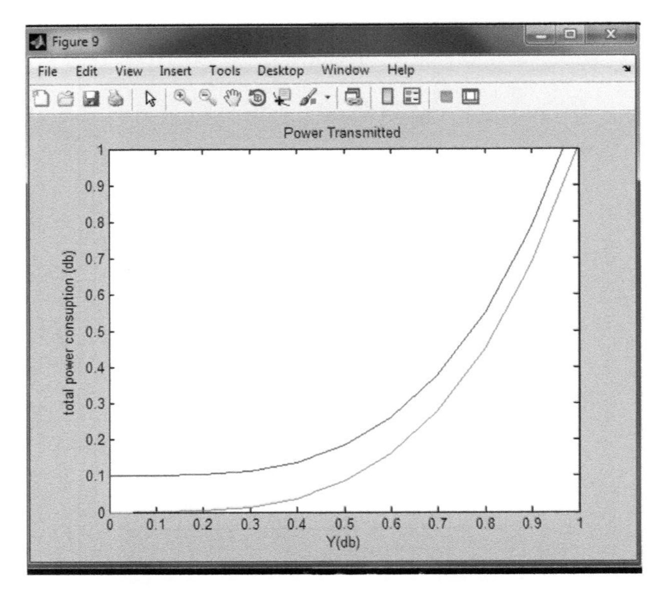

Figure 4.11 Illustrates that graphing which shows the require SINR for various source transmit power Ps yields vs total power consumption the desired result.

It responds to the Min-Max MSE to equalization with one relay and the Min-sum MSE, as well as the equalizer methodology for two relays.

4.6 Conclusion

Cooperative summarization of relay precoders and sources, as well as prefixed receivers, is researched in MIME networks. This is taken into account in order to maximize the worst stream SINR while maintaining a specified QoS based on relay power restrictions and source, as well as to reduce overall source and relay powers. The two non–convex optimization problems are comparable, and iterative alternating strategies have been devised to solve both. The optimization of the relay precoders and source may be represented as SDR and SOCP considerations for both challenges. Previous simulation findings suggest that the receivers decreased performance loss when compared to non-prefixed receivers, in which the equalizer is produced using relay precoders and source. The proposed design provides a good mix of complexity and performance, making it suitable for high-complexity receiver systems. For effective ICI cancellation in two-way conjugate transmission OFDM systems, an adaptive receiver will be developed in the future. The phase-rotated coupled cancellation idea, which uses two separate phase repeats on only two receive pathways instead of the one phase rotation utilized by the phase-repositioned conjugate revocation strategy on the two transmit paths, has been employed to design such a receiver.

References

[1] Ai.W , Huang.A, and Zhang.S, "New results on Hermitian matrix rank-one decomposition," *Math. Program: Ser. A*, vol. 128, no. 1-2, pp. 253–283, Jun. 2011.
[2] Behbahani.A , Merched.R, and Eltawil A.M, "Optimizations of a MIMO relay network," *IEEE Trans. Signal Process.*, vol. 56, no. 10, pp. 5062–5073, Oct. 2008.
[3] Bezdek.J and Hathaway.R, "Some notes on alternating optimization," *Advances in Soft Computing–AFSS*, pp. 187–195, 2002.
[4] Chae C.B, Tang.T, R. Jr.W.H., and Cho. S, "MIMO relaying with linear processing for multiuser transmission in fixed relay networks," *IEEE Trans. Signal Process.*, vol. 56, no. 2, pp. 727–738, Feb. 2008.
[5] Fu.Y , Yang.L, Zhu.W.P, and Liu.C, "Optimum linear design of two hop MIMO relay networks with QoS requirements," *IEEE Trans. Signal Process.*, vol. 59, no. 5, pp. 2257–2269, May 2011.
[6] Guan.W, and Luo.H, "Joint MMSE transceiver design in nonregenerative MIMO relay systems," *IEEE Commun. Lett.*, vol. 12, no. 7, pp. 517–519, Jul. 2008.
[7] Huang.Y and Palomar.D.P, "Rank-constrained separable semidefinite programming with applications to optimal beamforming," *IEEE Trans. Signal Process.*, vol. 58, no. 2, pp. 664–678, Feb. 2010.
[8] Hammerstom.I and Wittneben.A, "Power allocation schemes for amplify-and-forward MIMO-OFDM relay links," *IEEE Trans. Wireless Commun.*, vol. 6, no. 8, pp. 2798–2802, Aug. 2007.
[9] Havary-Nassab.V, Shahbazpanahi.S, and Grami.A, "Joint receiver transmit beamforming for multi-antenna relaying schemes," *IEEE Trans. Signal Process.*, vol. 58, no. 9, pp. 4966–4972, Sep. 2010.
[10] Huang.Y and Zhang.S, "Complex matrix decomposition and quadratic programming," *Math. Oper. Res.*, vol. 32, no. 3, pp. 758–768, Aug. 2007.
[11] Hoymann.C , Chen.W, Montojo.J, Golitschek.A, Koutsimanis.C, and Shen.X, "Relaying operation in 3GPP LTE: challenges and solutions," *IEEE Commun. Mag.*, vol. 50, no. 2, pp. 156–162, Feb. 2012.
[12] Joham.M, Utschick.W, and Nossek.J.A, "Linear transmit processing in MIMO communications systems," *IEEE Trans. Signal Process.*, vol. 53, no. 8, pp. 2700–2712, Aug. 2005.

[13] Laneman.J.N, Tse.D.N.C, and Wornell.G.W, "Cooperative diversity in wireless networks: efficient protocols and outage behavior," *IEEE Trans. Inf. Theory*, vol. 50, no. 12, pp. 3062–3080, Dec. 2004.

[14] Mohammadi.J , Gao.F, and Rong.Y, "Design of amplify and forward MIMO relay networks with QoS constraint," in *Proc. 2010 IEEE Global Communications Conference.*

[15] Munoz-Medina.O , Vidal.J, and Agustın.A, "Linear transceiver design in nonregenerative relays with channel state information," *IEEE Trans. Signal Process.*, vol. 55, no. 6, pp. 2593–2604, Jun. 2007.

[16] Palomar.D.P, and Eldar.Y.C, *Convex Optimization in Signal Processing and Communications*. Cambridge University Press, 2009.

[17] Rong.Y and Hua.Y, "Optimality of diagonalization of multi-hop MIMO relays," *IEEE Trans. Wireless Commun.*, vol. 8, no. 12, pp. 6068–6077, Dec. 2009.

[18] Rong.Y , "Optimal linear non-regenerative multi-hop MIMO relays with MMSE-DFE receiver at the destination," *IEEE Trans. Wireless Commun.*, vol. 9, no. 7, pp. 2268–2279, Jul. 2010.

[19] Rong.Y ,Tang.X, and Hua.Y, "A unified framework for optimizing linear non-regenerative multicarrier MIMO relay communication systems," *IEEE Trans. Signal Process.*, vol. 57, no. 12, pp. 4837–4851, Dec. 2009.

[20] Sanguinetti.L, D'Amico.A.A, and Rong.Y, "A tutorial on the optimization of amplify-and-forward MIMO relay systems," *IEEE J. Sel. Areas Commun.*, vol. 30, no. 8, pp. 1331–1346, Sep. 2012.

[21] Sendonaris.A , Erkip.E , and Aazhang.B , "User cooperation diversity—part I: system description," *IEEE Trans. Commun.*, vol. 51, no. 11, pp. 1927–1938, Nov. 2003.

Chapter 5

VLSI Implementation on MIMO Structure Using Modified Sphere Decoding Algorithms

R. Kabilan[1], R. Ravi[2], S. Shargunam[3], and R. Mallika Pandeeswari[4]

[1]Associate Professor, Department of ECE, Francis Xavier Engineering College, Tirunelveli, India
[2]Professor, Department of CSE, Francis Xavier Engineering College, Tirunelveli, India
[3]Full Time Ph.D Scholar, Dept of ECE, Francis Xavier Engineering College, Tirunelveli, India
[4]Full Time Ph.D Scholar, Dept of ECE, Francis Xavier Engineering College, Tirunelveli, India
Email: rkabilan13@gmail.com; fxhodcse@gmail.com; shargunamguna@gmail.com; mallikapandeeswari123@gmail.com

Abstract

Multiple antennas are used in both the transmitter and receiver ends of MIMO systems to increase spectrum efficiency. The implementation of two methods of decoding algorithms, the Viterbo-Boutros (VB) as well as Schnorr-Euchner (SE), is proposed in this method. The decoding algorithm is divided into a single FPGA device using a software/hardware co-design methodology. To maximize the decoding efficiency, three methods of parallelism are being adopted: using a concurrent channel matrix, using preprocessing and parallel decoding of imaginary or real selection, and concurrent execution of many phases in the closest lattice point search. With a Xilinx FPGA chip, the decoders for a 4x4 MIMO system with 16-QAM modulation have been prototyped. The VB and SE algorithms have hardware prototypes that relate to the support data rates of up to 84.4 and 38.2 Mb/s at a 22 dB SNR, which is faster than their respective implementations in a DSP.

5.1 Introduction

VLSI is the field that involves packing more and more logic devices into smaller and smaller areas. Circuits that would have previously taken up a lot of space can now be crammed into a mall pace a few millimeters across [6,7]. This has opened up a great opportunity to do things that were not possible before. Digital VLSI circuits are predominantly CMOS-based. The way normal blocks like latches and gates are implemented is different from what students have seen so far, but the behavior remains the same. All this miniaturization involves new things to consider. A lot of thought has to go into actual implementation as well as design. Let us look at some of the factors involved [10].

In a MIMO system, the total transmitted signal is spitting into multiple spatial paths, driving the capacity, so it will increase the spectral efficiency [1]. Most MIMO communication will have feedback with it.

Future wireless communication aims at higher data rates. Since the radio spectrum is limited, the requirement for high spectrum efficiency can be fulfilled by exploiting the spatial dimension of the radio channel [9]. This MIMO method brings a relevant increase not only in capacity but also in spectral efficiency, coverage, and reliability [2-4].

Spatial multiplexing is a very effective approach for boosting channel capacity while maintaining a high signal-to-noise ratio (SNR). At low signal-to-noise ratios (SNRs), MIMO wireless communication capacity rises linearly, whereas at high SNRs, capacity increases logarithmically [5].

When there is no channel information at the transmitter, diversity coding techniques are used. Diversity methods broadcast a single stream, then the signal is coded with the space-time coding method [8]. The signal is sent out from each of the transmit antennas, utilizing complete or near-orthogonal coding methods. To improve signal diversity, diversity takes advantage of the independent fading of various antenna links.

5.2 Proposed Methodology

The SoC technique is used to simplify the sphere decoder's design and increase its efficiency. The difficult preprocessing operations, like inversions and matrix factorizations, which will be in the embedded processor, the real-time decoding functions are divided into hardware module approaches. Both algorithms are suitable for MIMO systems. They are also called sphere decoders.

The VB and SE algorithms, which are two common sphere decoding techniques for MIMO detection, are summarised below. The research order within the lattice structure is the key distinction between these two techniques.

The VB method involves all methods of lattice points getting into a sphere on a lattice structure with a radius starting from the lower to upper bound, which is found in all layer searches.

5.2.1 VB decoding algorithm

It is the easiest method to locate the shortest lattice point. A metric update and a trace back are the two routines that make up the Viterbi algorithm.

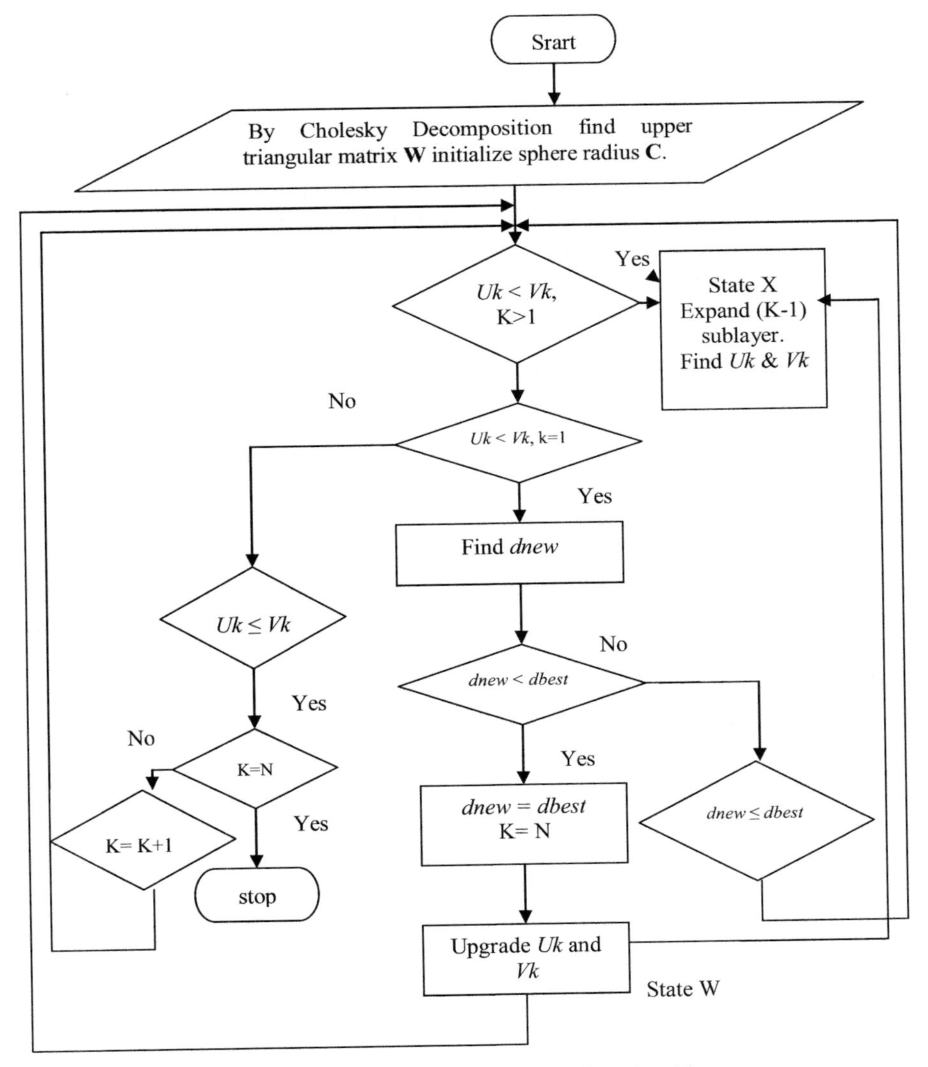

Figure 5.1 Flow chart for VB decoding algorithm.

Using a trellis diagram to describe state transitions, the metric update collects probabilities for all states based on the current input symbol. Once a path through the trellis has been discovered, the trace back algorithm reconstructs the original data. The studied layer's index is denoted by Uk, and the upper bound is denoted by Vk.

The following is a list of the VB algorithm's step-by-step procedures.

i) Reprocessing
The Cholesky decomposition algorithm is used to transform H into an upper triangular matrix W. Asymmetric, a positive-definite matrix is decomposed into a lower triangular matrix and a transpose of the lower triangular matrix using the Cholesky decomposition.

ii) Post-Production
A Finite-State Machine (FSM): A finite-state machine is a machine that has a finite number of states. An FSM is a behavioral model made up of a finite number of states, transitions between them, and actions. A finite state machine is a model of a machine with a primitive internal memory that can be abstracted.

5.2.2 SE decoding algorithm

The SE method is used for finding the closest lattice point that is close to the hyperplane. As a result, the SE algorithm does not require an initial radius. To find the shortest method of lattice point, it extends to find the distance between the receivers when it is used.

The dependency graph demonstrates how each state is related to the others. If the search starts at X and continues to Y and Z, Next, state X will naturally depend on both states of Y and Z. The next orthogonal distance is being measured in states Y or Z and it is being used in state X. The state X is also known as a self-evaluator. State Y has no data dependency on X since the mentioned states will have different dimensions if state Y follows X in the search process. This will be continued in State Z.

The SE algorithm has no bounding constraints. Each layer's bound does not need to be calculated and updated. As a result, time-consuming square root procedures are omitted. Second, by adopting the no-decreasing order of examination, the chances of recognizing the correct layer early are increased. Furthermore, because the lattice point search begins at the Elgar point, there is no need for an initial radius.

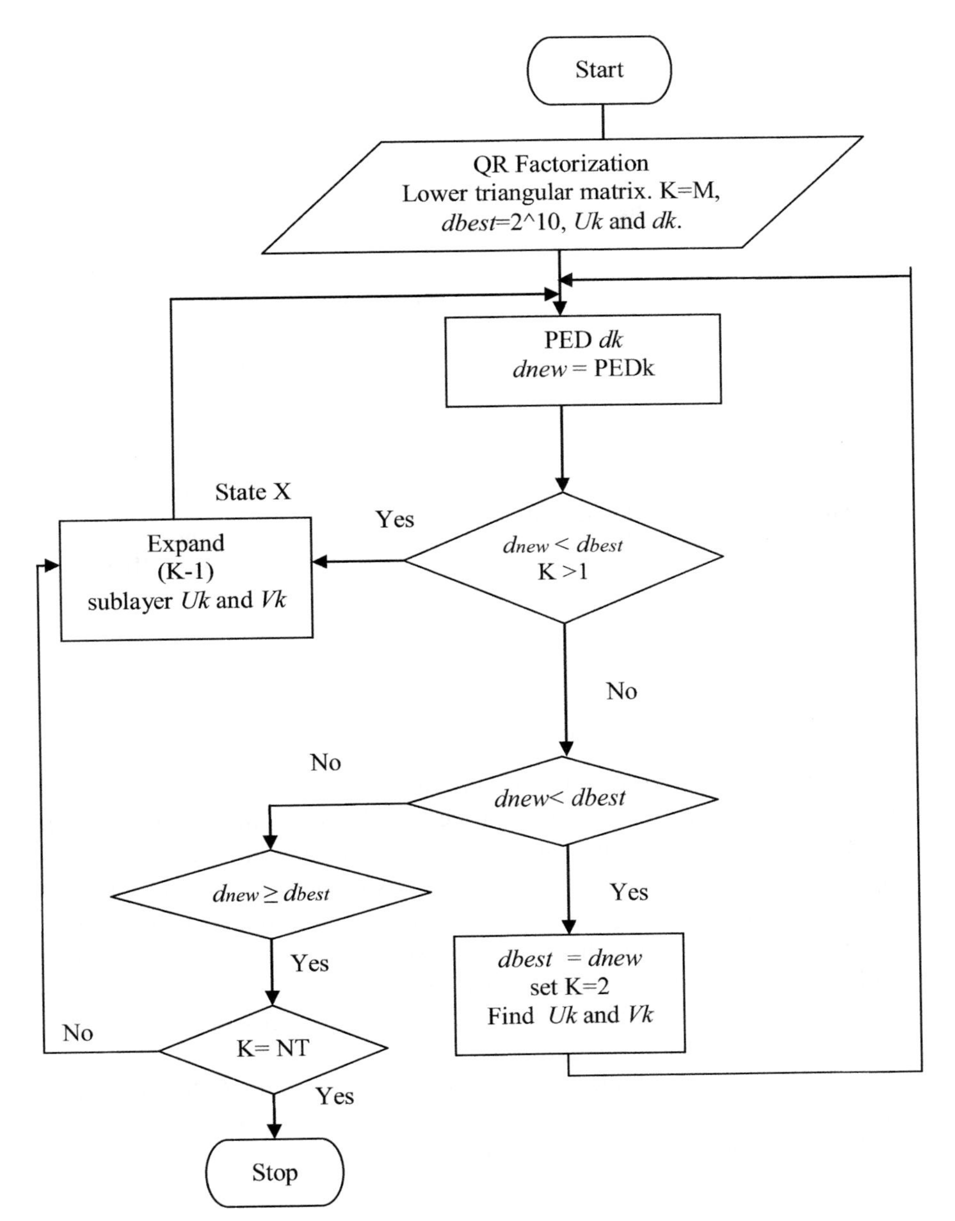

Figure 5.2 Flow chart for SE decoding algorithm.

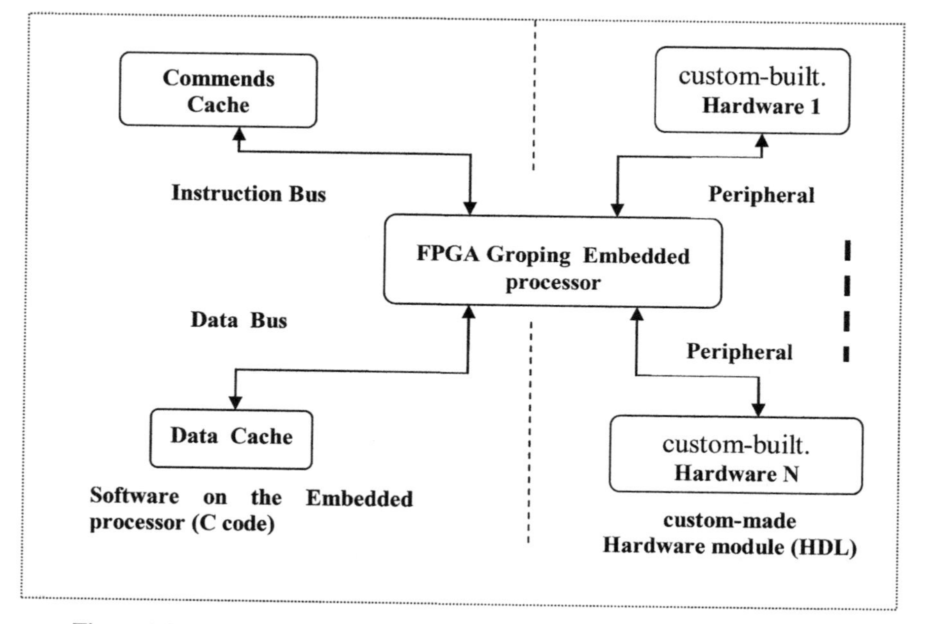

Figure 5.3 FPGA-based SoC architecture for sphere decoder implementations.

5.2.3 SOC architecture on FPGA

It is a generic SOC architecture that provides sphere decoding algorithms. It will enhance the nearer neighbor lattice search and reduction in design. It converts the lattice matrix into an upper or lower triangular matrix, which is being seen in VB techniques. This is known as preprocessing, and it entails matrix factorizations like QR decomposition, as well as matrix inversions.

The SoC architecture has an embedded CPU, specialized hardware modules, and a shared peripheral bus for the use of data transfer. The decoding procedure is then partitioned into each of the processor units using a hardware/software code sign approach.

The preprocessing findings were transmitted from the CPU to the specialized hardware devices on a regular basis. The FPGA-based SoC design provides increased programmability and customization.

5.3 Result and Conclusion

The proposed, Viterbo–Boutros (VB) sphere decoding algorithm is performed in order to search for the closest lattice point. It is explained and implemented using VHDL.

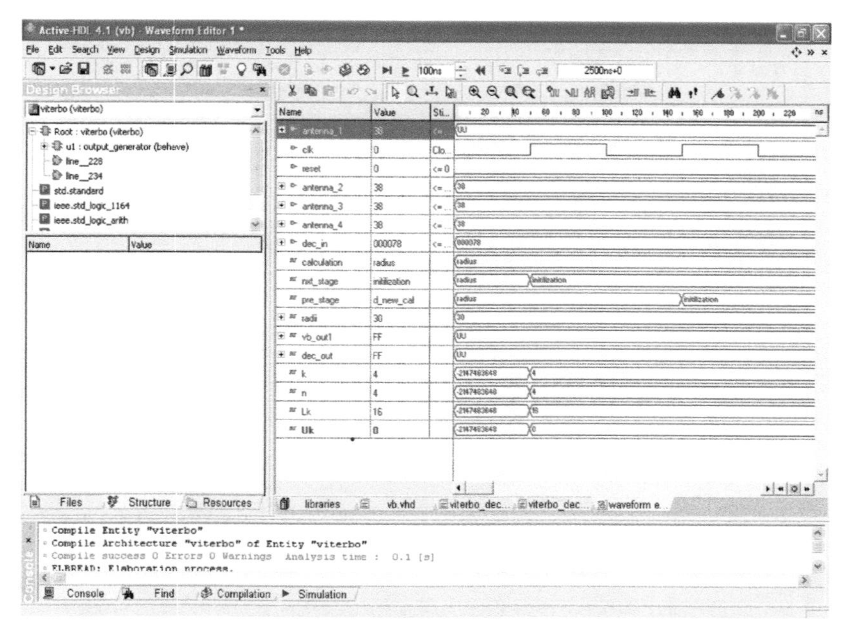

Figure 5.4 Simulated output of VB.

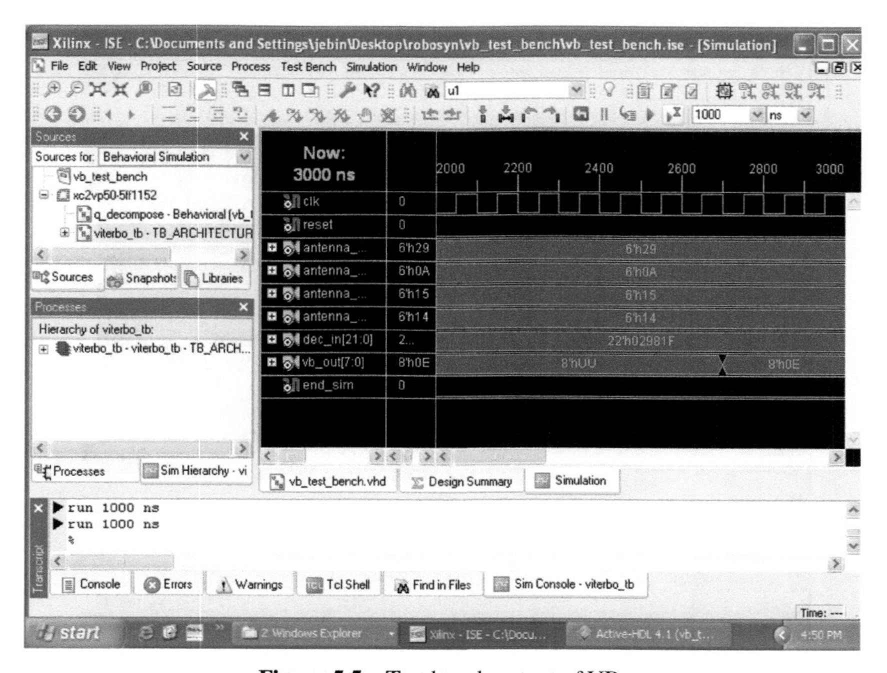

Figure 5.5 Test bench output of VB.

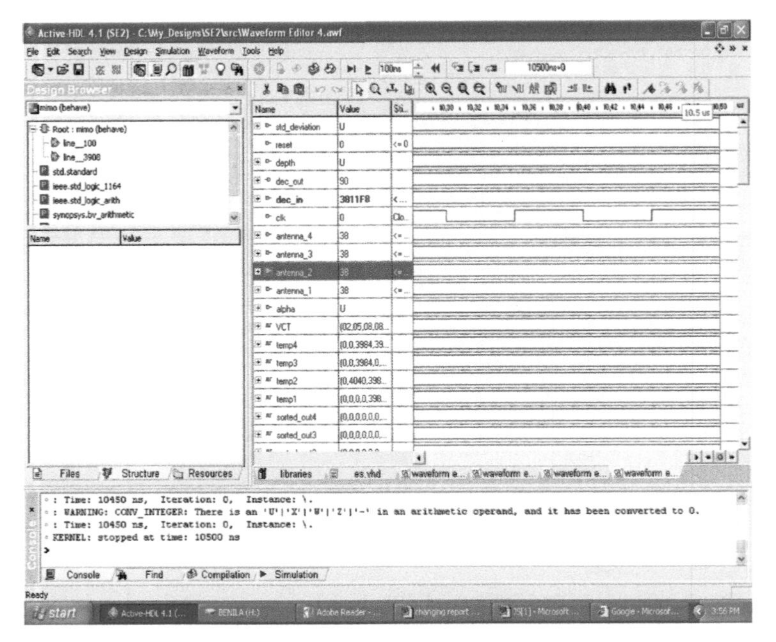

Figure 5.6 Simulated output of SE.

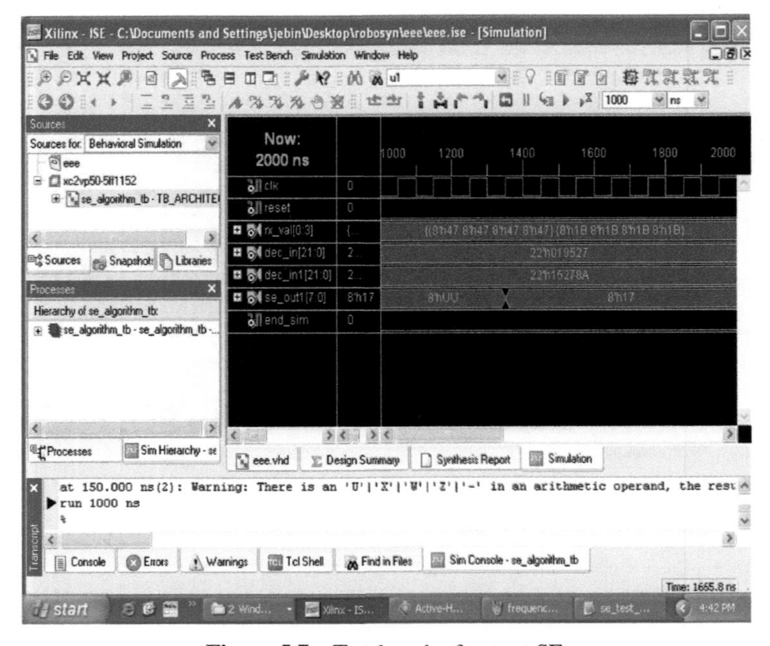

Figure 5.7 Test bench of output SE.

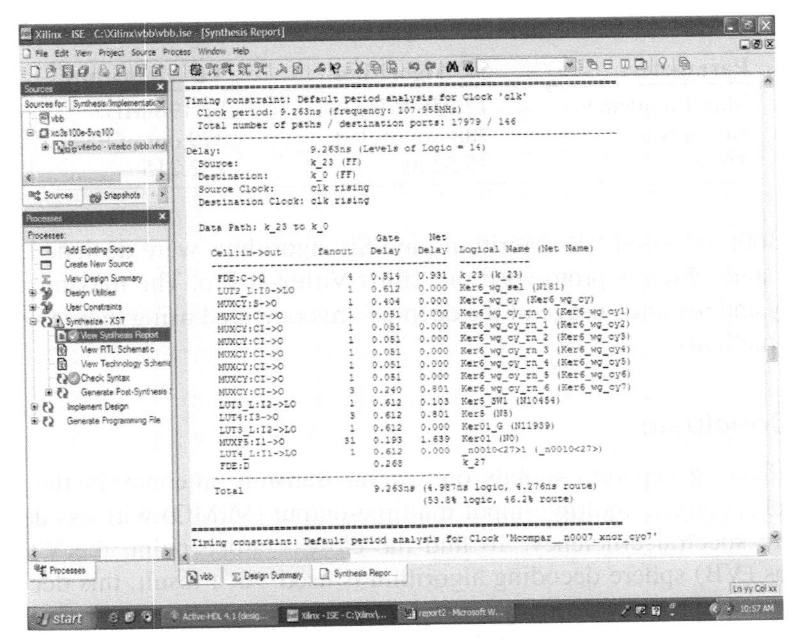

Figure 5.8 Synthesis report of VB.

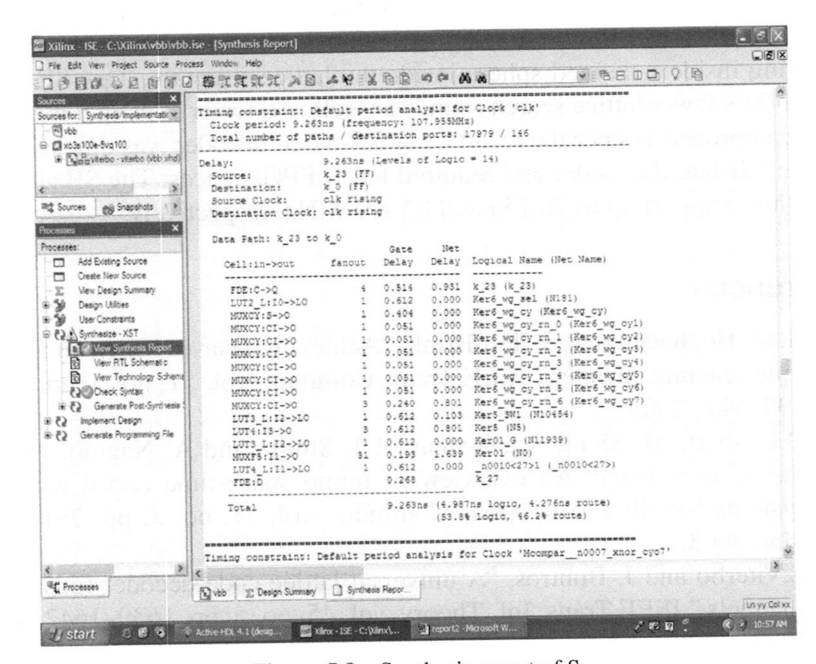

Figure 5.9 Synthesis report of S.

Table 5.1 Experimental result.

Parameters	SE Algorithm	VB Algorithm
Max-Frequency	77.256 MHz	83.198 MHz
No. of Slices	388 out of 960	379 out of 959
Time	35.52 ns	132.215

Both proposed VB algorithms and SE algorithms were designed using EDK3 and which is prototyped on Xilinx Virtex-II Pro. The maximum frequency and resource usage of the decoder are compared using these two proposed methods.

5.4 Conclusion

By delivering separate signals on several transmit antennas in the same channel spectrum, multiple-input multiple-output (MIMO) wireless devices improve spectral efficiency. To find the closest lattice point, the Viterbo–Boutros (VB) sphere decoding algorithm is used. As a result, this decoding technique improves efficiency. The platform architectures and implementations for two typical spherical decoding algorithms enabling MIMO detection are presented in this study. This approach examines each search layer first, from the lower limit to the higher bound, with all feasible lattice points occurring inside a defined sphere inside the lattice structure. This SE algorithm offers fewer lattice search rounds as well as a lower construct than just the VB approach. As per the results, our SE-based decoder was quicker than just the VB-based decoder and required fewer FPGA slices. The SE and VB algorithms support up to 76.153 and 82.173 MHz, respectively.

References

[1] B.M. Hochwald and S. Ten Brink, “Achieving near-capacity on a multiple-antenna channel,” IEEE Trans. Commun, vol. 51, no. 3, pp. 389–399, Mar. 2003.

[2] D.Gesbert, M. Shafi, S. Da-shan, P. J. Smith, and A. Naguib, “From theory to practice: An overview of mimo space-time coded wireless systems,” IEEE J. Sel. Areas Commun., vol. 21, no. 3, pp. 281–302, Mar. 2003.

[3] E.Viterbo and J. Boutros, “A universal lattice code decoder for fading channels,” IEEE Trans. Inf. Theory, vol. 45, no. 5, pp. 1639–1642, May 1999.

[4] E.Agrell, T. Eriksson, A. Vardy, and K. Zeger, "Closest point search in lattices," IEEE Trans. Inf. Theory, vol. 48, no. 8, pp. 2201–2214, Aug. 2002.

[5] A.Adjoudani, E. C. Beck, A. P. Burg, G. M. Djuknic, T. G. Gvoth, D. Haessig, S. Manji, M. A. Milbrodt, M. Rupp, D. Samardzija, A. B. Siegel, I. Sizer, T. , C. Tran, S. Walker, S. A. Wilkus, and P. W. Wolniansky, "Prototype experience for MIMO blast over third-generation wireless system," IEEE J. Sel. Areas Commun., vol. 21, no. 3, pp.440–451, Mar. 2003.

[6] D.Garrett, L. Davis, S. ten Brink, B. Hochwald, and G. Knagge, "Silicon complexity for maximum likelihood MIMO detection using spherical decoding," IEEE J. Solid-State Circuits, vol. 39, no. 9, pp. 1544–1552, Sep. 2004.

[7] R. Kabilan., Ravi R., Rajakumar G., Esther Leethiya Rani S., Mini Minar V. C. 'An Improved VLSI architecture using wavelet filter bank for blur image applications', ARPN Journal of Engineering and Applied Sciences, Vol. 10, no. 6, April 2015.

[8] C. Selsi Aulvina., R. Kabilan, 'LOW power and area efficient borrow save adder design', Proceedings of the International Conference on Smart Systems and Inventive Technology, ICSSIT 2018, 2018, pp. 339–342, 8748832.

[9] G. Prince Devaraj., Dr. R. Kabilan., J. Zahariya Gabriel., U. Muthuraman., N. Muthukumaran., R. Swetha, 'Design and Analysis of Modified Pre-Charge Sensing Circuit for STT-MRAM', IEEE Xplore, Proceedings of the 3rd International Conference on Intelligent Communication Technologies and Virtual Mobile Networks, ICICV 2021, 2021, pp. 507–511.

[10] A.Burg, M. Borgmann, M. Wenk, M. Zellweger, W. Fichtner, and H. Bolcskei, "VLSI implementation of MIMO detection using the sphere decoding algorithm," IEEE J. Solid-State Circuits, vol. 40, no. 7, pp. 1566–1577, Jul. 2005.

Chapter 6

Overcrowding Cell Interference Detection and Mitigation in a Multiple Networking Environment

R. Ravi[1], R. Mallika Pandeeswari[2], R. Kabilan[3], and S. Shargunam[4]

[1]Professor, Department of CSE, Francis Xavier Engineering College, Tirunelveli, India
[2]Full Time Ph.D Scholar, Dept of ECE, Francis Xavier Engineering College, Tirunelveli, India
[3]Associate Professor, Department of ECE, Francis Xavier Engineering College, Tirunelveli, India
[4]Full Time Ph.D Scholar, Dept of ECE, Francis Xavier Engineering College, Tirunelveli, India
Email: fxhodcse@gmail.com; mallikapandeeswari123@gmail.com; rkabilan13@gmail.com ;shargunamguna@gmail.com

Abstract

Heterogeneous networking seems to be a viable option for extending wireless coverage, reducing congested bandwidth, and improving system performance in next-generation wireless networks. Furthermore, interference between overlay cells is one of the most difficult issues to overcome when deploying heterogeneous networks. Short-period interference is created by femtocell uplink broadcasts involving macrocell users. This is shown by the fact that femtocell networks suffer significant performance degradation when utilizing OFDMA or SC-FDMA protocols that lack information on the interferer channels that link macrocells and femtocells to the base station. Consider employing precoded FDMA (P-FDMA) transmission to novel detectors to prevent shot interference. The PFDMA strategy may maximize overall signal intensity in any receive antenna and throughput for many receive antenna systems by using a multi-stream transmission. Our suggested structure accumulates multipath variety in the presence of interference to the knowledge of interferer channels, according to the findings, and our concepts seem to be dependable on incursion power levels. Simulation evidence backs up our theoretical results..

6.1 Introduction

Wireless communication [4] is the transmission of information between two sites that are not connected by an electrical line. Cells make up a wireless access network. Thanks to a restored transmitter that connects with the base station BS [11], each cell is operational. The majority of users connect to the network using a dedicated resource. Cells are utilized to increase coverage and extend it [7]. A BSC is a device that links BSs to the main network. Access control, cell handover, and other radio resource allocation activities are handled by BSC. A wireless cellular network is seen in Figure 6.1.

In the direct meaning, ICT is being used to reduce its own energy consumption; in the indirect sense, ICT is being used to reduce carbon emissions; and in the systematic sense, ICT is partnering with other sectors of the economy to simply deliver energy efficiency. The influence of ICT, especially data centers, on worldwide CO2 emissions is enormous [3]. Furthermore, power costs make up a significant component of the OPEX. The issues and

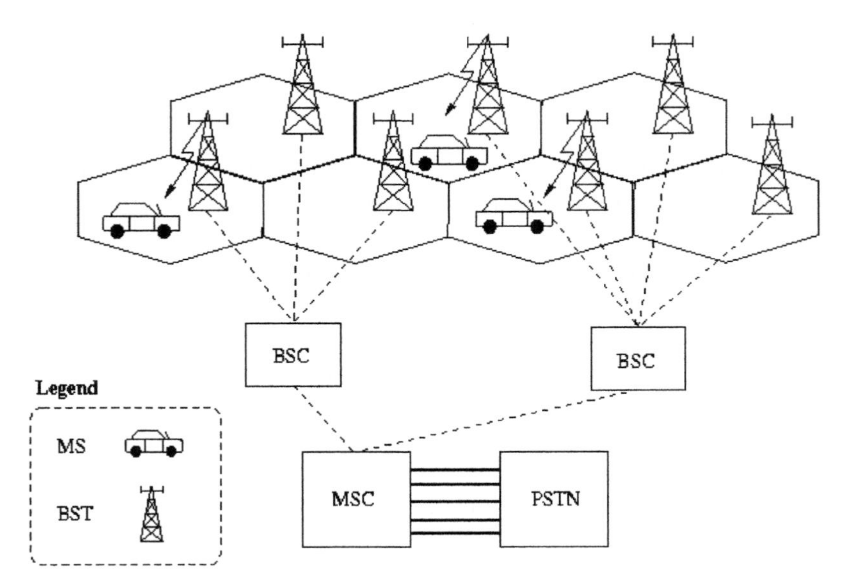

Figure 6.1 Wireless cellular networks [from internet].

problems that must be addressed in order to reduce carbon emissions and power prices in the industry are discussed in this research [6].

Femtocell network

The implementation of femtocells is a cost-effective solution to the problem of communications infrastructure for micro-inaction in cellular networks [10]. Users buy and install a simple low-base station in their townhouse. Femtocell can cover an area with a radius of 5 to 20 meters. As like macrocell femto cell will operate with the same kind of licensed spectrum [8].

Heterogeneous network

A heterogeneous network connects computer systems that use different systems and/or protocols. For example, LANs that attach Microsoft Windows and Linux-based personal computers to Apple Macintosh computers, for example, are heterogeneous [5].

Orthogonal frequency-division multiple access (OFDMA)

OFDMA is the most widely used version of the OFDM virtual modulation scheme. OFDMA allows for low-data-rate transmission to multiple users [1].

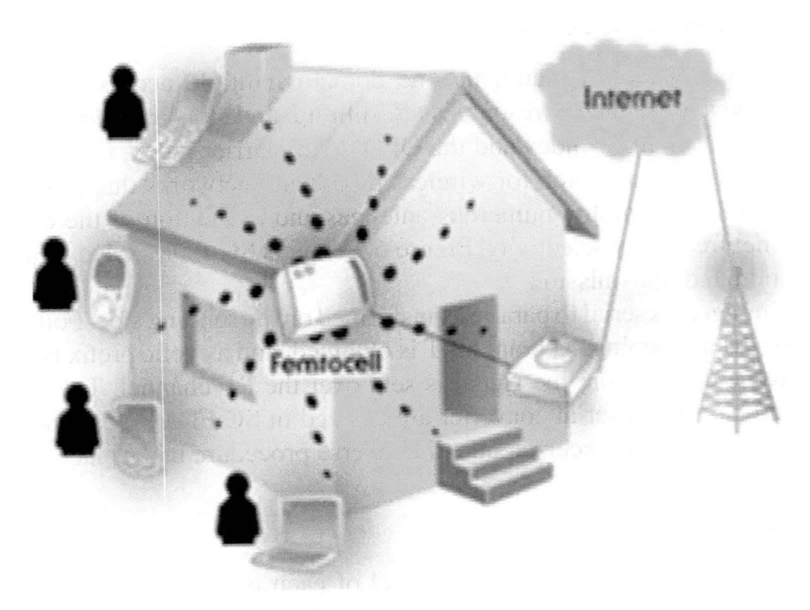

Figure 6.2 Femtocell.

Single-carrier frequency division multiple access (SC-FDMA)
SC-FDMA will be regarded as the most enticing channel access technique for 3GPP LTE uplink. SC-FDMA is indeed a mixture of FDMA and a single SC-FDE with a structure and performance similar to OFDMA [9]. When compared to multi-carrier transmissions such as OFDMA, this single-carrier transmission has a lower PAPR [2].

6.2 Proposed System

Traditional macrocell networks are under tremendous pressure as a result of the huge expansion in wireless indoor communications. FBSs were carefully built to allow low-mobility users to attain high-performance inside communications in order to solve the problem of indoor coverage. Femtocells are recommended in current cellular radio standards, such as LTE, the Global Scheme for Mobile Communications, and Update Accordingly for Microwave Access, since they boost the network's wireless capacity, decrease power consumption, and save money in the operation.

This suggested J-DET achieves performance equivalent to P-FDMA systems with subcarrier information even without a comprehension of tried-to-interfere subcarriers. The results of the analysis are shown in the figure below.

6.3 OFDMA and SCFDMA

OFDMA is a frequency domain/time domain multiple access hybrids in which time and frequency resources are split up, and slots are allotted based on the OFDM symbol index and the OFDM sub-carrier index. OFDMA has been seen as a great match for wireless broadband networks due to its scalability, ability to employ numerous antennas and evaluation of the efficacy of channel frequency selectivity. Figure 6.3 depicts SC-FDMA and OFDMA generation and transmission.

The signal is serial to parallel transformed on the sending side. Following the subcarrier mapping, M point DFT is obtained, and a cyclic prefix is added to decrease error before the signal is sent over the RF channel. The n point DFT and parallel to serial converter are utilized in SC-FDMA, as illustrated in the picture. On the receiver side, the inverse procedure takes place.

6.4 Results and Discussion

In order to find out the interference level of each user BER-SNR , BLER-SNR and SE-SNR performance is analyzed and evaluated.

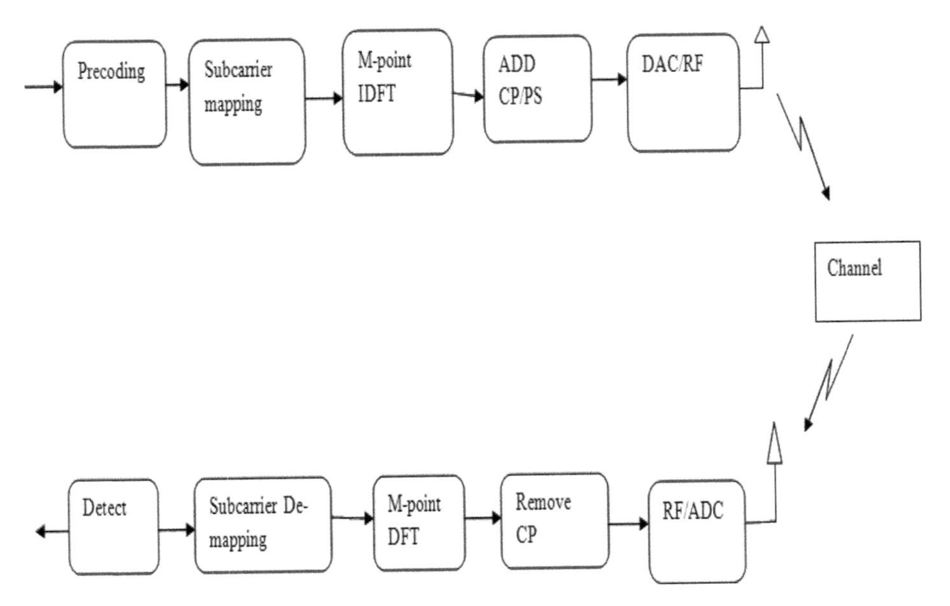

Figure 6.3 SCFDMA and OFDMA transmission and reception.

6.4.1 BER–SNR graph of two users

The figure 6.4 indicate the BER-SNR graph of two users. BER-SNR performance is compared for two users which use the 4QAM modulation scheme to indicate the interference level of each user. Figure 6.4 (a) indicates the less interfered user while figure 6.4 (b) indicates the highly interfered user. Comparing both with the same SNR range BER will be higher for the second user. Thus the interference will increase as the distance increases.

6.4.2 BLER–SNR graph of two user

The figure 6.5 indicate the BLER-SNR graph of two users. Block error versus signal to noise ratio performance is compared for two users which use the 4QAM modulation scheme to indicate the interference level of each user. Figure 6.5(a) indicates the less interfered user while figure 6.5 (b) indicates highly interfered user comparing both for the same SNR range BLER will be lower for the second user. Thus the interference will increase as the distance increased.

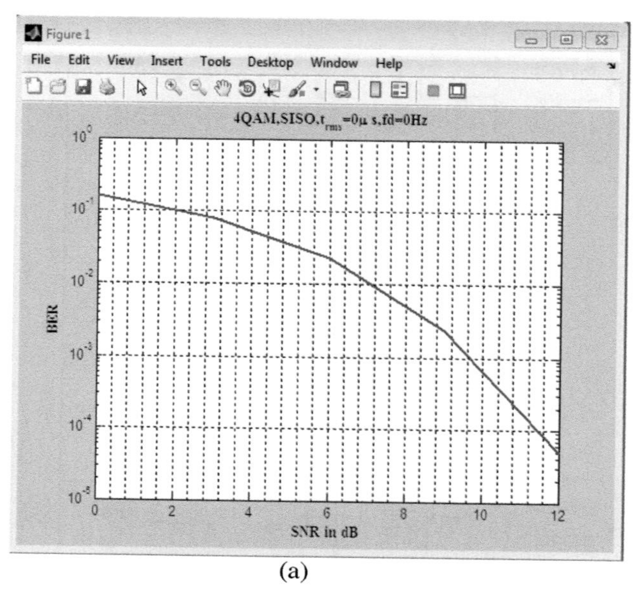

(a)

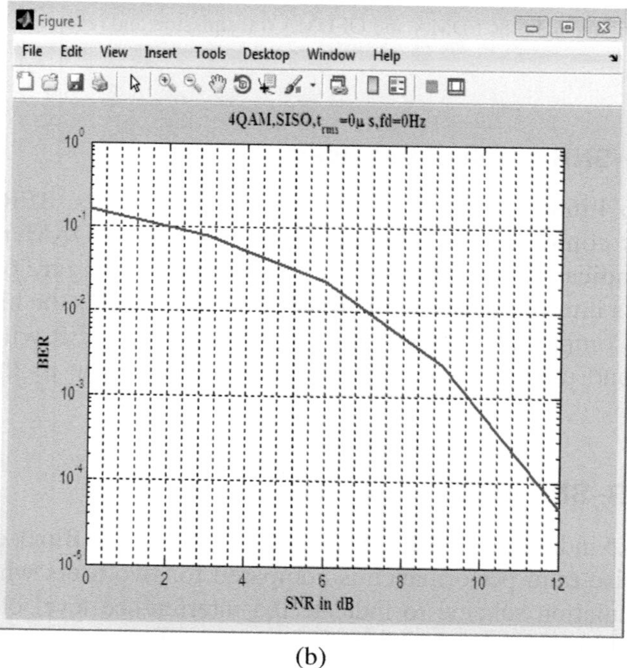

(b)

Figure 6.4 (a), (b) BER–SNR graph of two users.

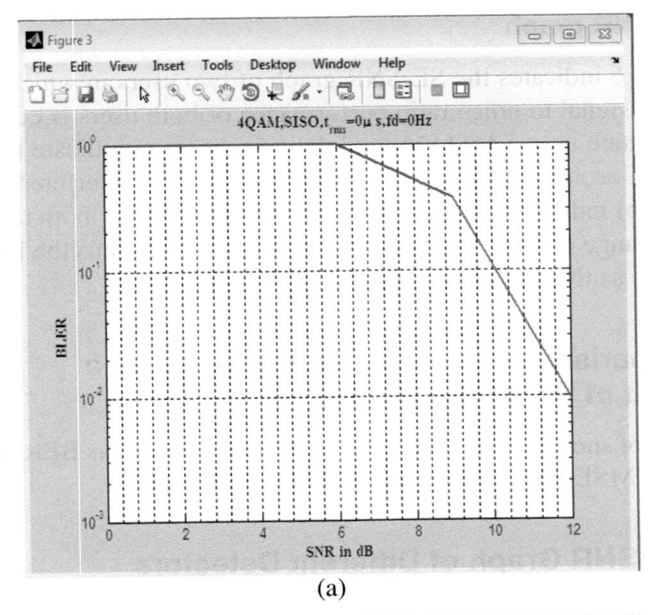

(a)

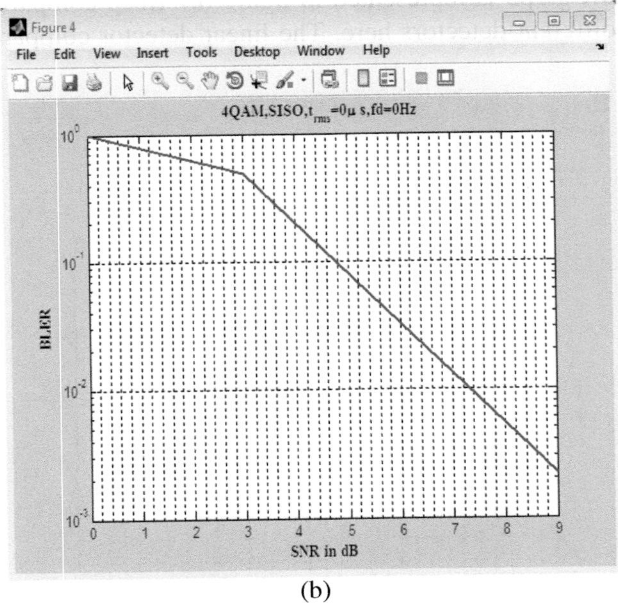

(b)

Figure 6.5 (a), (b) BLER–SNR graph of two uses.

6.4.3 SE-SNR graph

The figure 6.6 indicates the SE-SNR graph of two users in which the single error versus signal to noice ratio performance of both users is compared for two users which uses a 16QAM modulation scheme to indicate the interference level of each user. Figure 6.6 (a) indicate the less interfered user while Figure 6.6 (b) indicates highly interfered user. Comparing both these for the same SNR range will be SA lower for the second user. Thus the interference will increase as the distance increases.

6.5 Comparison of Detector Performance as a Result of Shot Interference

In presence of short interference, Like performances such as BER-SNR AND CHANNEL MSE-SNR of various detectors are compared.

6.6 BER–SNR Graph of Different Detectors

The BER-SNR graph is represented in figure 6.7 Keep comparing the contribution of different detectors here. The linear detector compares the BER

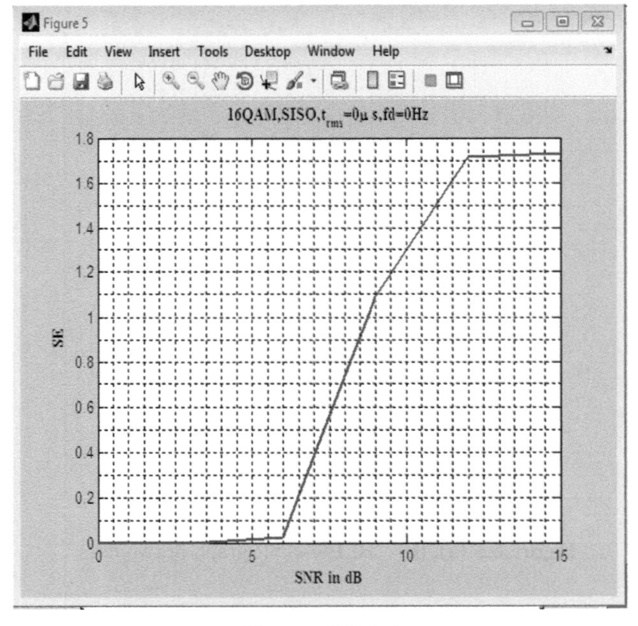

Figure 6.6 (a)

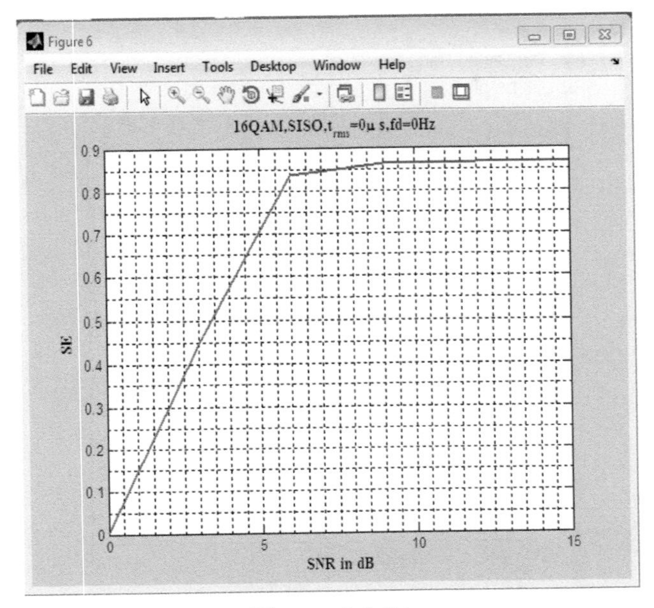

Figure 6.6 (b)

Figure 6.6 (a), (b) E-SNR graph of two users.

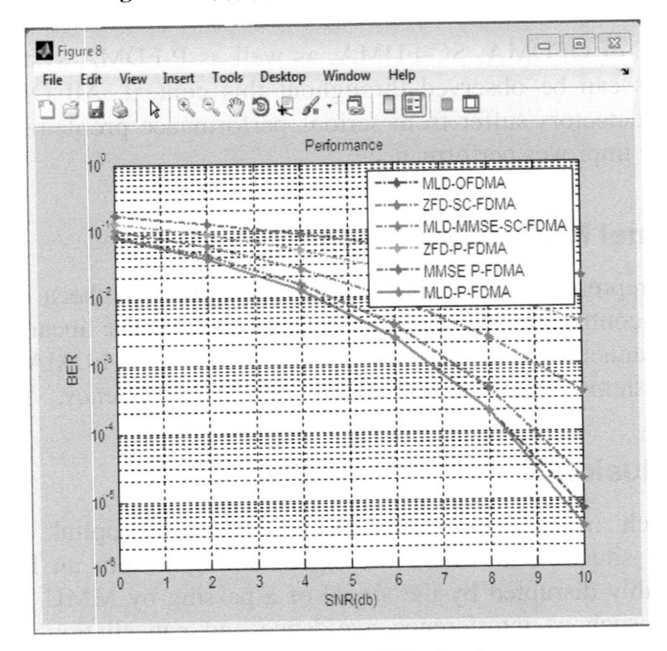

Figure 6.7 BER–SNR Graph.

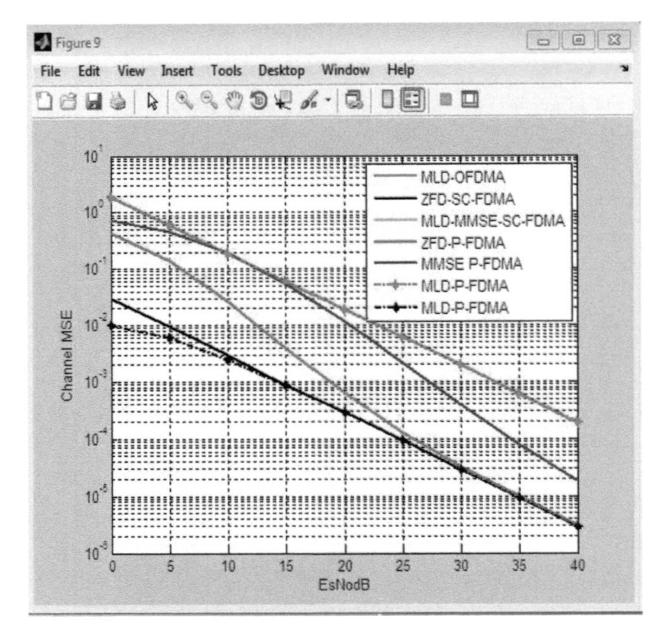

Figure 6.8 Channel MSE –ESN0 Graph.

performance of OFDMA, SC-FDMA, as well as P-FDMA systems. Some many trends can be observed throughout this context. All OFDMA and SC-FDMA detectors suffer from serious performance problems. P-FDMA dramatically improves performance.

6.7 Channel MSE–ESN0 Graph

Figure 6.8 represents the channel MSE-E_SN_0 graph. Make a comparison between the contributions of different detectors below. The linear sensor will sense the channel MSE vs. performance of OFDMA, SC-FDMA, as well as P-FDMA systems. P-FDMA dramatically improves efficiency.

6.8 Conclusion

This research investigated a macrocell-to-femtocell uplink co-channel involvement situation in which the signal strength from an FMU at an FBS is forcibly disrupted by the signal of a passing-by MMU. Traditional approaches, such as interference avoidance and cancellation, confront a number of obstacles, including accurate channel prediction, valid subcarrier allocation data for all MMUs, the expense of feedback between the FMU and the FBS, and so on. As a result, they proposed a method in which the interference was provided as impulsive noise in this circumstance. Simulations were also utilized to show that the P-FDMA system is reliable at different degrees of SIRs and the number of interfered subcarriers. Shot interference: the suggested design may be an effective and acceptable solution for femtocell systems encountering shot interference, taking into consideration the limited system requirements as well as the high throughput of our researched systems. In the long run, a booster is used to pick the same low-bandwidth network, decreasing interference.

References

[1] Qi Zhou, Xiaoli Ma, Shot Interference Detection and Mitigation for Heterogeneous Networks. IEEE Transactions On Vehicular Technology, Vol. 63, No. 1, January 2014

[2] D. Lopez-Perez, A. Valcarce, G. de la Roche, and J. Zhang, "OFDMAfemtocells: A roadmap on interference avoidance," IEEE Commun. Mag.,vol. 47, no. 9, pp. 41–48, Sep. 2009.

[3] V. Chandrasekhar and J. Andrews, "Uplink capacity and interference avoidance for two-tier femtocell networks," IEEE Trans. Wireless Commun., vol. 8, no. 7, pp. 3498–3509, Jul. 2009.

[4] S. Kishore, L. J. Greenstein, H. V. Poor, and S. C. Schwartz, "Uplink user capacity in a multi cell cdma system with hotspot microcells," IEEE Trans. Wireless Commun., vol. 5, no. 6, pp. 1333–1342, Jun. 2006.

[5] M. E. Sahin, I. Guvenc, M.-R. Jeong, and H. Arslan, "Handling CCI and ICI in OFDMA femtocell networks through frequency scheduling," IEEE Trans. Consum. Electron., vol. 55, no. 4, pp. 1936–1944, Nov. 2009.

[6] Y. Tu, K.-C. Chen, and R. Prasad, "spectrum sensing of ofdma systems for cognitive radio networks," IEEE trans. Veh. Technol., Vol. 58, no. 7, pp. 3410–3425, sep. 2009.

[7] W. Yune, C.-H. Choi, G.-H. Im, J.-B. Lim, E.-S. Kim, Y.-C. Cheong, and K.-H. Kim, "SC-FDMA with iterative multiuser detection: improve mentson power/spectral efficiency," IEEE commun. Mag., Vol. 48, no. 3,pp. 164–171, mar. 2010.

[8] M. Yavuz, F. Meshkati, S. Nanda, A. Pokhariyal, N. Johnson, B. Raghothaman, and A. Richardson, "interference management and performance analysis of UMTS/HSPA+ femtocells," IEEE commun. Mag., Vol. 47, no. 9, pp. 102–109, sep. 2009.

[9] P. Hoeher, S. Badri-hoeher, W. Xu, and C. Krakowski, "single-antenna co-channel interference cancellation for TDMA cellular radio systems," IEEE wireless commun. Mag., Vol. 12, no. 2, pp. 30–37, apr. 2005

[10] P. Xia, S. Zhou, and G. B. Giannakis, "bandwidth- and power-efficient multicarrier multiple access," IEEE trans. Commun., Vol. 51, no. 11, pp. 1828–1837, nov. 2003.

[11] qi zhou, student member, ieee, and xiaoli ma, senior member ieee shot interference detection and mitigation for heterogeneous networks, ieee transactions on vehicular technology, vol. 63, no. 1, january 2014

Chapter 7

A Baseband Transceiver for MIMO-OFDMA in Spatial Multiplexing Using Modified V-BLAST Algorithm

R. Kabilan[1], R. Ravi[2], S. Shargunam[3], and R. Mallika Pandeeswari[4]

[1]Associate Professor, Department of ECE, Francis Xavier Engineering College, Tirunelveli, India
[2]Professor, Department of CSE, Francis Xavier Engineering College, Tirunelveli, India
[3]Full Time Ph.D Scholar, Dept of ECE, Francis Xavier Engineering College, Tirunelveli, India
[4]Full Time Ph.D Scholar, Dept of ECE, Francis Xavier Engineering College, Tirunelveli, India
Email: rkabilan13@gmail.com; fxhodcse@gmail.com; shargunamguna@gmail.com; mallikapandeeswari123@gmail.com

Abstract

The MIMO/OFDMA uplink baseband transceiver system is being proposed. It is an ICI-cancellation motivated CFO predictor that also interoperates with channel estimation as well as a MIMO detector to compensate for reducing interference caused by the carrier frequency. A lower compliance approach is also offered. With MIMO communication devices, VBLAST seems to be an effective detection method. The proposed modified VBLAST gives computational complexity to the algorithm it will reduce the total number of iterations. The complexity of both the detection and the signal is considerably reduced as a result of this simplification. An arrangement for compensating for CFOs at an OFDMA system's base station is also provided. The suggested fix is ideal for high-data-rate communications. In addition, a low-average implementation remedy with good performance is provided.

7.1 Introduction

OFDM seems to be a digital information encoding technique that makes use of multiple carrier frequencies. OfDM has become a widely known wideband digital system, regardless of wireless, including over copper wires, which is used in application areas including 4G mobile communications, mobile as well as television systems [9-10].

The OFDM spread spectrum method distributes data with a large number of carriers at a large number of maximum spacing at accurate frequencies [4]. Because multipath channels seem to be common throughout terrestrial broadcasting scenarios, this is useful. Even though alternate iterations of a signal interfere with others, it becomes extremely difficult to extract the original information. OFDM is another name for multi-carrier modulation as well as discrete multi-tone modulation. It really is the modulation innovation being used on digital television in many countries [11,15].

7.1.1 OFDM modulation

IFFT is being used to generate OFDM symbols. The spectrum of both the subcarriers is closely spaced as well as merged to obtain higher bandwidth efficiency. Nulls in the spectrum of each sub-carrier land in the center of all the other sub-carriers [5].

7.1.2 FDMA

FDMA allows users to allocate more bandwidths, or channels, to themselves. Like some other multiple access systems, FDMA coordinates access between

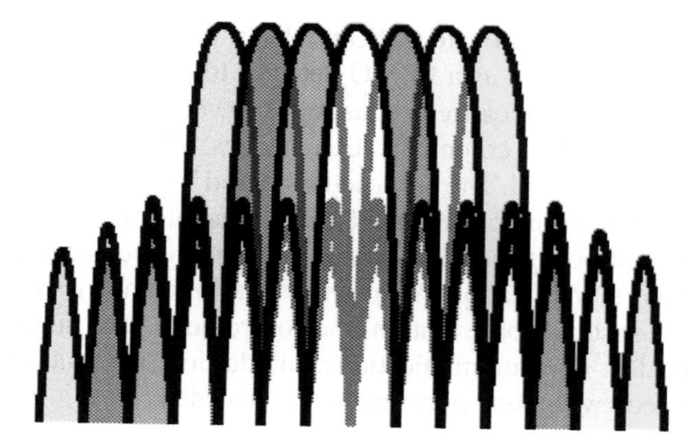

Figure 7.1 OFDM Modulation subcarrier allocation diagram [from Internet].

many different users [7]. It is, in fact, extremely common throughout the satellite. FDMA is a channel access method and is used as a channel allocation protocol throughout multiple-access protocols [14].

7.1.3 OFDMA

One of the most frequent methods of OFDM is OFDMA, which would be a multi-user variation of such popular orthogonal OFDM. As seen in the graphic below, OFDMA offers various access to individual users. Various allow users to send low-bandwidth data in real-time [13].

7.1.4 MIMO OFDM

MIMO/OFDM is a very well & broadly applied technology. Wireless technology uses plenty of antennas to send as well to receive radio signals. This will enable service providers to construct a broadband type wireless system [12].

The MIMO system utilizes several antennas to simultaneously broadcast data in small parts towards the receiver, which can process and assemble the data flows. This technique, known as spatial multiplexing, increases data transmission speed depending on the total number of antennas [8]. Moreover, because all data is sent in the same frequency band but with distinct spatial fingerprints, this method uses the spectrum more efficiently [2,6].

7.2 Existing Method

To handle the multiuser transmission, an OFDMA system combines the FDMA protocol and the OFDM technology. As a result, compared to

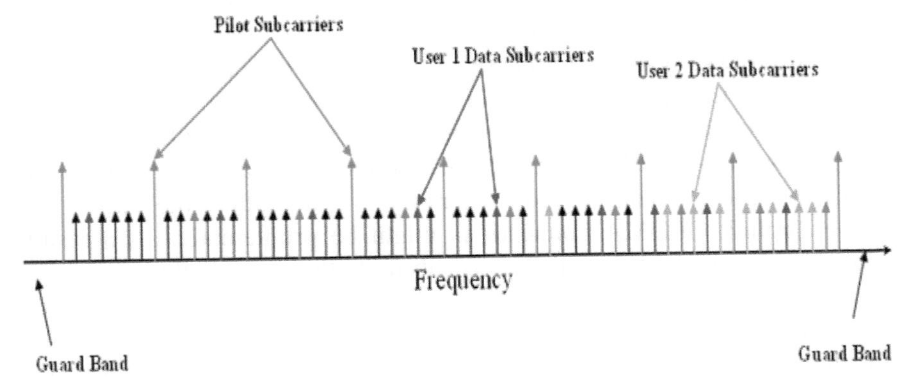

Figure 7.2 Carrier allocation for OFDMA signal [from Internet].

traditional OFDM systems, OFDMA will provide flexibility as well as transmission efficiency in the system. Normally OFDMA is based on OFDM, it is extremely susceptible to synchronization and channel estimate errors. Downlink OFDMA synchronization has been the subject of numerous investigations [1]. Uplink OFDMA communications, on the other hand, have issues when numerous users try to send data at the same time.

7.2.1 Synchronization algorithms for MIMO OFDMA systems

The suggested methodology is mainly worried about uplink asynchronous transmission protocols because it is meant to be used in a Base Station setting. Determining frequency offset, time synchronization, and other channel parameters for each user is a particularly difficult task because of the enormous number of unknown variables. We propose a method for breaking a multidimensional maximizing text into sub-problems individual maximizing problems based on EM. The reliability of this estimator as well as its capacity to endure the near-far effect are clearly demonstrated by the simulation results.

7.3 MIMO Transceiver

This section will look at a 2x2 MIMO-OFDMA model in spatial multiplexing mode which depends upon the IEEE 802.16e standard. This method provides the receiver will be divided into two groups based on its way of operating [3].

7.4 Proposed Method

Multiuser transmission is supported by an OFDMA system, which combines the FDMA protocol with the OFDM technology. As a result, compared to traditional OFDM systems, OFDMA gives you more control over channel bandwidth and transmission power. OFDMA is very susceptible to synchronization and channel estimate errors because it is based on OFDM's basic principle. Downlink OFDMA synchronization has been addressed in a number of researches. When numerous users try to send data at the same time, however, uplink OFDMA connections might be problematic. As a result, uplink transmission synchronization is substantially more challenging.

To design the transceiver in low complexity and low synchronization errors and to improve the overall performance, this system is proposed. ICI Cancellation based CFO Estimation, V-Blast Detector in MIMO Decoder, Synchronization Algorithm used is Ranging Process is used in this system.

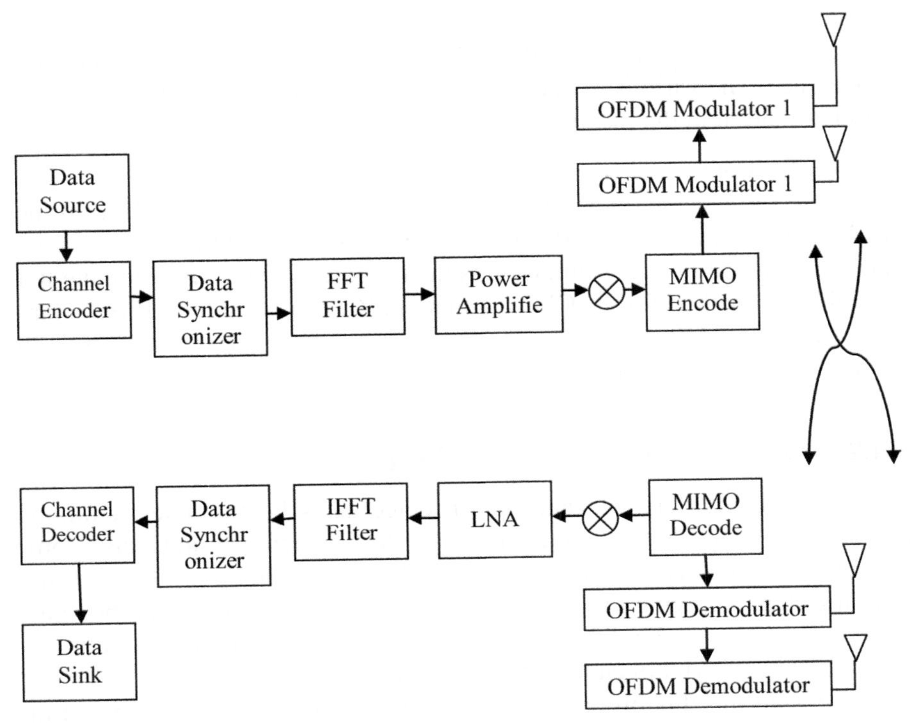

Figure 7.3 Block diagram of transceiver.

7.4.1 Module description

Transmitter Section:

Data sources generate message data, that one would like to communicate from transmitter to receiver section through MIMO Channel. Message data from the Data sources are fed to the channel encoder, and the binary input is fed as the input to the output section. Basically, a Transceiver is designed using VCO and PLL circuits to keep the phases or frequencies matched.

MIMO Encoder is indeed a methodology for sending diverse data from the source to the destination and making a way that can be easily distinguished. The OFDMA Modulator receives it after that. OFDMA is made up of two protocols: OFDM and FDMA. Here is a design for a 2x2 MIMO transceiver. As a result, two OFDMA modulators and demodulators are used. OFDMA is extremely susceptible to synchronization and channel estimation errors.

MIMO Channel:

The signals passing through the MIMO channel will suffer from many-interferences and various frequency offsets. It is removed using the ICI based CFO Estimation in the receiver section.

Receiver Section:

Here the reverse operation is done to recover the original data. In the MIMO Decoder, the V-Blast decoder is used to help detect the received data. The synchronization algorithm used is the ranging process. By using the ICI Cancellation method for the CFO estimation, ICI caused in the MIMO channel is get reduced. By designing the transceiver with these modules, we can achieve low synchronization errors.

7.4.2 Proposed modified V-BLAST algorithm

In successive interference cancellation detection schemes, later iterations have a lesser cancellation effects because the interference and noise components have been canceled during the early stage of iterations through the ordering step. Applying this property to V-BLAST detection it can be interpreted as the bigger the difference of the norm an early iteration, the lesser the efficiency of symbol cancellation in later iterations

Considering this property, a modified version of V BLAST based on stopping the iterations if its effect of cancellation is sufficiently small, simplifying the calculation complexity. According to the ordering process, in every iteration stage, the ordinary V-BLAST algorithm selects the smallest norm of the remaining norm values and performs the cancellation process. However, our proposed modified VBLAST algorithm compares the selected smallest norm value with the average of the remaining norm values. If the selected norm value is smaller than the average norm value multiplied by a constant parameter C, it proceeds with the cancellation process. It means the selected one is much smaller than the others and it is well worth being canceled.

7.5 Result and Discussion

The figure 7.5 shows the Maximum likelihood of Frequency offset as well as Timing offset. In this, ML estimation between time as well as frequency offset, first the parameters are declared. With these given parameters, OFDM symbols are generated in the transmitter part. Next in the channel section AWGN is added and then it is given to the receiver part. There in the

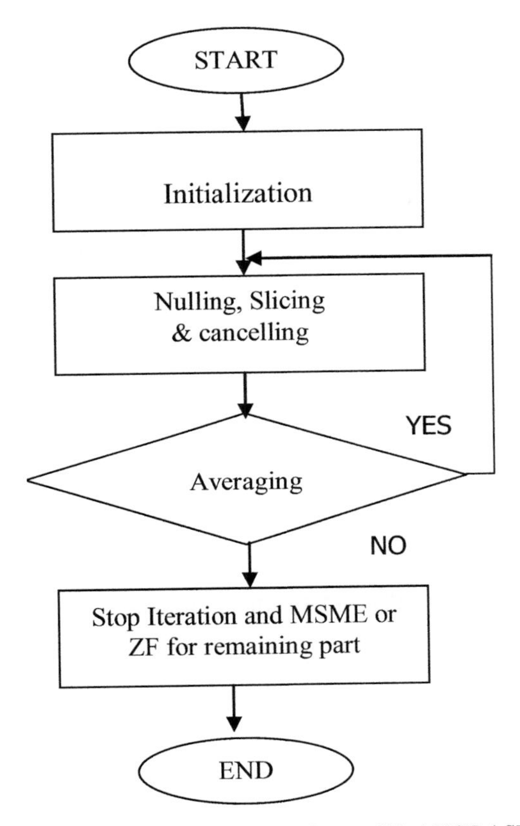

Figure 7.4 Propose algorithm for modified V BLAST.

receiver part, ML estimation of time and frequency offset is done and the results are displayed. By this, the performance of the system can be easily analyzed.

This figure 7.6 shows the Bit error Rate versus Eb/No It also shows the various plots when we are changing the number of users in the receivers in the Rayleigh channel.

This figure 7.7 illustrate between bit error rate vs Average Eb/No of Equalization with Successive Interference Cancellation – With the Modified VBLAST algorithm. This shows the Bit error rate for Binay Phase Shift Keying modulation towards Rayleigh fading method along with two transmitters as well as 2 receiver MIMO channel systems.

This Figure 7.8 shows the 2*2 MIMO –OFDMA simulation. It also gives the BER vs SNR graph. By this, the performance of the system can be easily analyzed

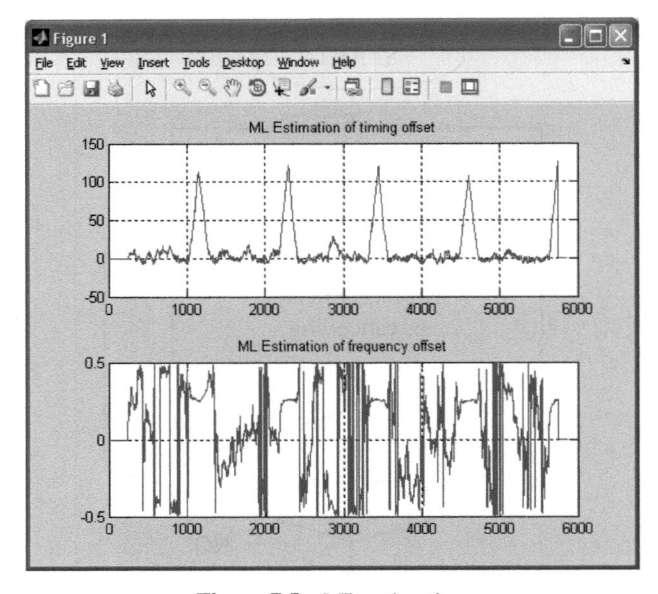

Figure 7.5 ML estimation.

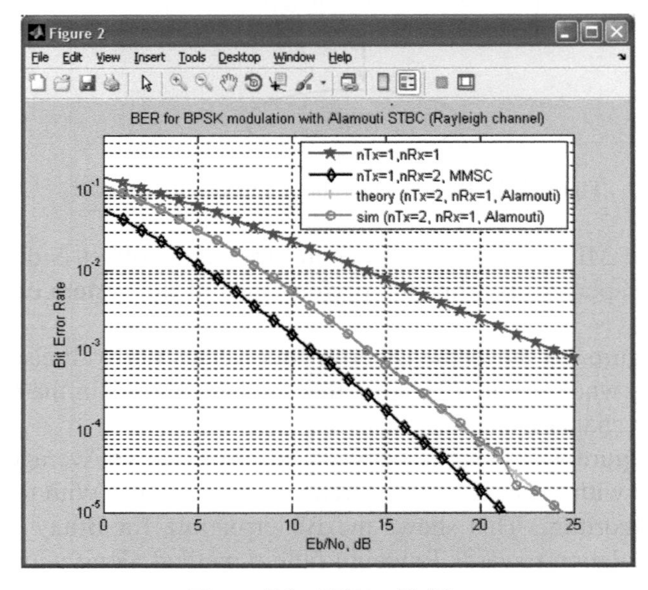

Figure 7.6 BER vs Eb/No.

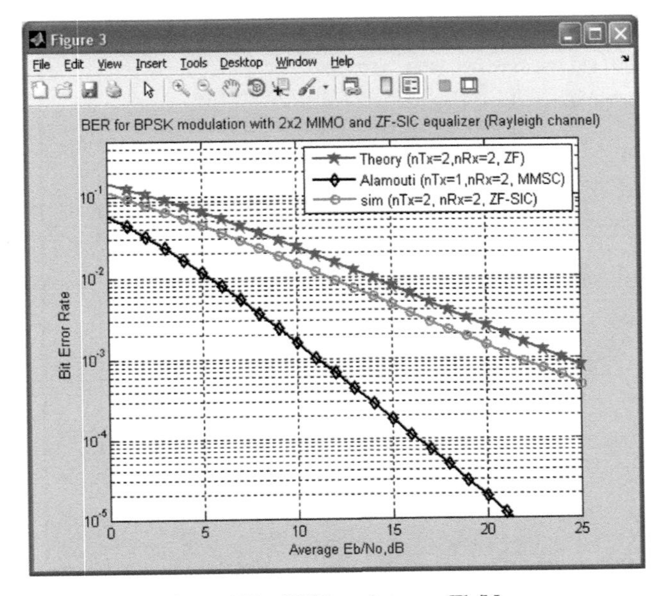

Figure 7.7 BER vs Average Eb/No.

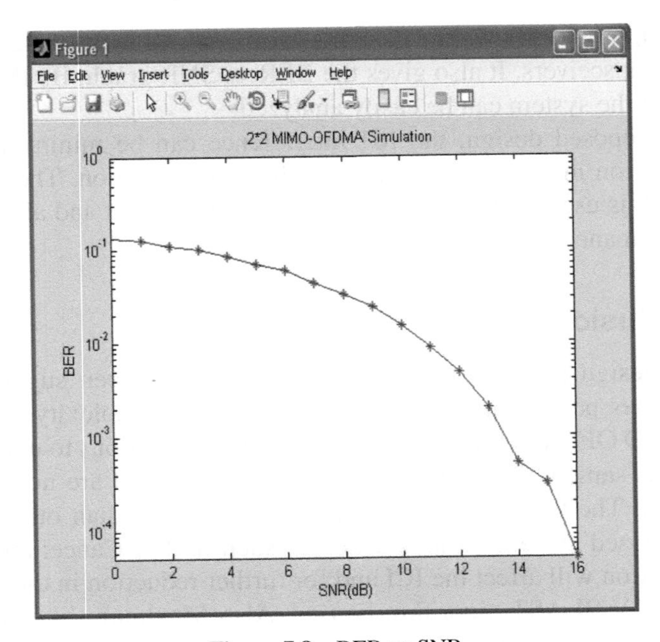

Figure 7.8 BER vs SNR.

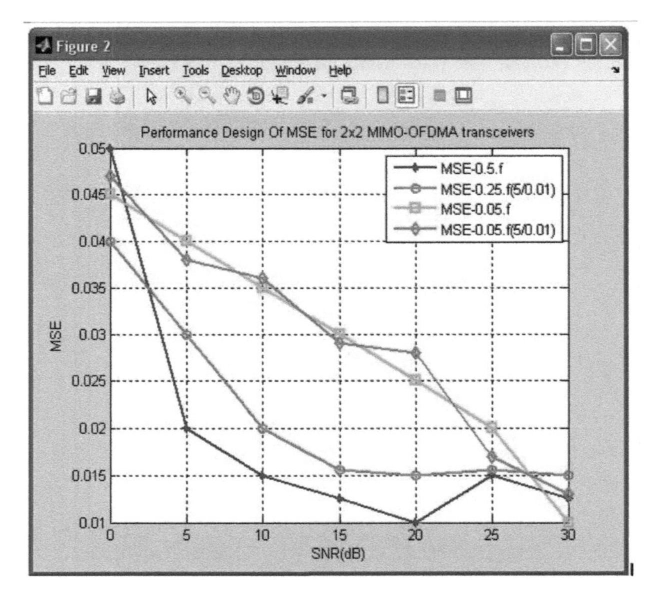

Figure 7.9 MSE vs SNR.

This figure 7.9 shows the design performance of MSE for 2*2 MIMO OFDMA Transceivers. It also gives the BER vs SNR graph. By this the performance of the system can be easily analyzed.

The proposed design, the ICI interference can be minimized by the CFO estimation in conjunction with the channel estimation. The CFO estimator which is used in the proposed design is very robust and also provides better performance.

7.6 Conclusion

A system design related to spatial multiplexing has been suggested that either includes performance assessment as well as complexity analysis of such a MIMO OFDMA transceiver. The essential limitations to uplink transmission, this same CFO, and also the multipath channel are now all taken into account. The CFO estimator is much more robust than other designs. In this proposed system various schemes such as ICI Cancellation based CFO estimation will affect the ICI and for further reduction in the complexity Modified V-BLAST algorithm is used. Also Maximum likelihood estimation is needed to find time and frequency offset for the synchronization process. By using this type of estimation in the early stage, they can easily synchronize by using these offset values. Moreover, if all of the elements are electronically accessible as well as inexpensive, a low-average complexity implementation remedy that achieves high reliability is provided. Moreover, the suggested transceiver has been shown to have improved performance. As a result, this same proposed solution lends itself to high-data communications.

References

[1] F. Xiaoyu and M. Hlaing, "Initial uplink synchronization and power control (ranging process) for OFDMA systems," in Proc. IEEE GLOBECOM, vol. 6, 2004, pp. 3999–4003.

[2] Z. Cao, U. Tureli, and Y. D. Yao, "Deterministic multiuser carrierfrequency offset estimation for interleaved OFDMA uplink," IEEE Trans. Commun., vol. 52, pp. 1585–1594, Sept. 2004.

[3] M.-O. Pun, M. Morelli, and C.-C. J. Kuo, "Maximum-likelihood synchronization and channel estimation for OFDMA uplink transmissions," IEEE Trans. Commun., vol. 54, no. 4, pp. 726 736, Apr. 2006.

[4] S. Barbarossa, M. Pompili, and G. B. Giannakis, "Channel-independent synchronization of orthogonal frequency-division multiple access systems," IEEE J. Select. Areas Commun., vol.20, pp. 474–486, Feb. 2002.

[5] Y. Yingwei and G. B. Giannakis, "Blind carrier frequency offset estimation in SISO, MIMO, and multiuser OFDM systems," IEEE Trans. Commun., vol. 53, no. 1, pp. 173–183, Jan. 2005.

[6] J. M. Lin and H. P. Ma, "A baseband transceiver for IEEE 802.16e-2005 MIMO-OFDMA uplink communications," in Proc. IEEE GLOBECOM, Nov. 2007, pp. 4291–4295.

[7] 802.16e: IEEE Standard for Local and Metropolitan Area Networks Part 16: Air Interface for Fixed and Mobile BroadbandWireless Access Systems Amendment for Physical and Medium Access Control Layers for Combined Fixed and Mobile Operation in Licensed Bands and Corrigendum 1, IEEE Std. 802.16e-2005, Dec. 2005.

[8] G. Schrack, W. Wu, and X. Liu, "A fast distance approximation algorithm for encoded quadtree locations," in Proc. Canadian Conference, vol. 2, Sept. 1993, pp. 1135–1138.

[9] C. W. Yu and H. P. Ma, "A low complexity scalable MIMO detector," in Proc. IWCMC 2006, Vancouver, Canada, July 2006, pp. 605–610.

[10] Technical Specification Group Radio Access Network: User Equipment (UE) radio transmission and reception (FDD) (Release 7), 3GPP TS 25.101 v7.0.0, June 2005.

[11] G. J. Foschini and M. J. Gans, "On limits of wireless communications in a fading environment when using multiple antennas," Wireless Personal Commun., pp. 311–335, June 1998.

[12] G. D. Golden, J. G. Foschini, R. A. Valenzuela, and P. W. Wolniansky, "Detection algorithm and initial laboratory results using V-BLAST space-time communication architecture," Electron. Lett., vol. 35, pp. 14–15, Jan. 1999.

[13] I. E. Telatar, "Capacity of multi-antenna Gaussian channels," Eur. Trans. Telecommun., vol. 10, pp. 585–595, Nov.–Dec. 1999.

[14] W. J. Choi, R. Negi and J. M. Cioffi, "Combined ML and DFE decoding for the V-BLAST system," in Proc. ICC, pp. 1243–1248, June 2000.

[15] J. A. C. Bingham, "Multicarrier modulation for data transmission: an idea whose time has come," IEEE Trans. Commun., vol. 28, pp. 5–14.

Chapter 8

Hardware Implementation of OFDM Transceiver Using Simulink Blocks for MIMO Systems

R. Ravi[1], J. Zahariya Gabriel[2], R. Kabilan[3], and R. Mallika Pandeeswari[4]

[1]Professor, Department of CSE, Francis Xavier Engineering College, Tirunelveli, India
[2]Associate Professor, Department of ECE, Francis Xavier Engineering College, Tirunelveli, India
[3]Associate Professor, Department of ECE, Francis Xavier Engineering College, Tirunelveli, India
[4]Full Time Ph.D Scholar, Dept of ECE, Francis Xavier Engineering College, Tirunelveli, India
Email: fxhodcse@gmail.com; zahagabs@gmail.com; rkabilan13@gmail.com; mallikapandeeswari123@gmail.com

Abstract

In wireless communication applications, the MIMO-OFDM system plays a vital role. The multiple inputs and multiple outputs components of the MIMO process were the focal points. The communication process aims to reduce energy consumption at the original transmission input signal level while also improving the effectiveness of wireless communication applications. The hardware-based VLSI architecture will be used to modify this technology. This architecture is designed to improve the performance of the transceivers in 802.11 MIMO-OFDM systems. This system uses Matlab-simulink software and a hardware FPGA board to construct a 4*4 MIMO-OFDM architecture. Matlab simulation is used to convert VHDL code and verify the output for the VLSI simulation graph. To optimize the transmission timing in the MIMO architecture process, design the blocks in the simulink unit and alter the block layout in the overall OFDM-MIMO unit, as well as build the sub-system functions. The hardware architecture for the Simulink 802.11 MIMO-OFDM system design is implemented using the system generator and Xilinx software. The design technique for optimizing the hardware program design process for MIMO encode and decode processes, as well as transceiver operations, using VHDL code. The goal of the simulation is to organize the MIMO-OFDM blocks and increase the energy efficiency of wireless communication systems.

8.1 Introduction

MIMO techniques have emerged as potential characteristics in wireless communications in recent years, allowing for increased spectral efficiency or better service quality [19]. Successful deployments of current FPGA-based reconfigurable computers require adequate runtime operating system support. Previous attempts to create operating systems for FPGA-based devices have mostly concentrated on the challenge of hardware task scheduling. Modern FPGA-based systems, on the other hand, necessitate features such as Internet connectivity, file system access, home network integration, and sophisticated user interface mechanisms in addition to basic job scheduling [2]. MIMO systems use numerous antennas at the transmitter and receiver to leverage the spatial dimension in addition to the time and frequency dimensions, resulting in excellent spectral efficiency [20]. MIMO technology has been the technique of choice in many wireless standards because of its great spectral efficiency [1].

The mathematical manipulation of an information stream to change or improve it in any way is known as digital signal processing (DSP). It is defined as the representation of discrete-time, discrete frequency, or other discrete domain signals by a series of numbers or symbols, as well as the processing of these signals [18]. The key themes for system development are hardware and software technology [3]. A CPU and numerous DSPs may be

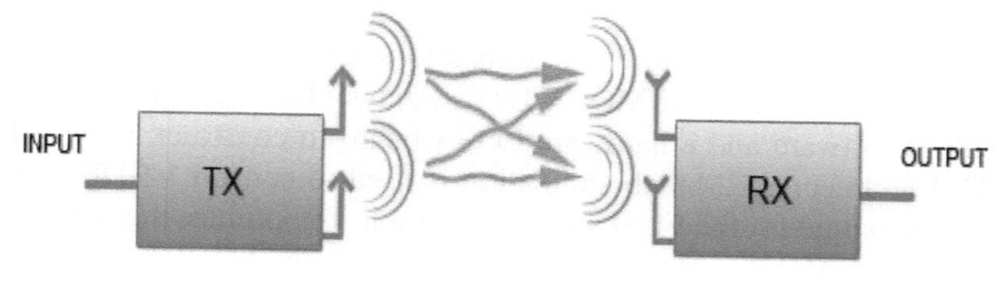

Figure 8.1 MIMO systems [from internet].

used to perform almost all speech recognition. Various architectures and tools that facilitate application development have been established in tandem with the growing areas of use of voice-processing technology [17, 4].

One of the most significant technological advancements in modem communication is digital communication using MIMO systems. Time diversity, frequency diversity, and spatial variety are all examples of At the transmitter or receiver end, spatial diversity necessitates the employment of numerous antennas [12–16].

8.2 Existing System

Under the premise of statistical independence among the source signals, ICA can estimate a demixing matrix and separate signals. The majority of ICA algorithms take advantage of non-Gaussianity. However, because of the time-consuming nonparametric estimate of the score functions and the sluggish convergence of the gradient algorithm used to solve the estimating equations, this method has a significant computational cost [5-7]. The hardware results reveal a thorough examination of RTL schematics and the Test Bench. The software simulation results in this research show a 2dB reduction in peak levels [8,10].

8.2.1 Fast ICA

The Fast ICA algorithm is a fixed-point type approach for independent component analysis and blind source separation that is both computationally efficient and resilient.

8.2.2 Efficient variant of fast ICA algorithm (EFICA)

Fast ICA's most efficient version is based on the following observations: For different sources, the symmetric Fast ICA algorithm can be run with varied nonlinearities. It is feasible to incorporate auxiliary constants in the symmetrization phase of each iteration, which can be tweaked to reduce the mean square estimation error in one row of the estimated de-mixing matrix [9].

8.2.3 Sphere decoding algorithm

A new architecture for efficient VLSI implementation of the spherical decoding algorithm for communication systems is presented, as well as a description of an actual implementation of a 44-decoder for 16-QAM modulation.

So far, just one ASIC implementation of the spherical method has been documented, to our knowledge. In high-performance MIMO communication systems, maximum probability detection is critical [11].

8.3 Proposed System

The suggested approach aims to reduce the amount of energy used in data transmission in wireless communication systems. This system is suitable for a wide range of wireless communication applications. The application was primarily concerned with the data transmission unit's energy consumption. The purpose of the MIMO-OFDM block is to optimize the block and subsystem layout for the proposed system methodology and to shorten the transceiver architecture process for the MIMO-OFDM architecture. Simulink software converts VHDL code and implements hardware. VLSI simulation in Xilinx software architecture is the suggested system architecture design.

8.4 MIMO-OFDM

In frequency selective fading conditions, MIMO in combination with OFDM is a viable technology for achieving high data rates and huge system capacity for wireless communication systems. Next-generation WLAN is the most likely application of MIMO.

8.5 Channel Estimation (CE)

The most significant benefits of combining MIMO and OFDM over a wireless link are the increased data rates and system capacity. To get better data rates, we require precise CSI on the receiver. Signals must be transmitted from separate antennas at the same time in the case of OFDM with multiple antennas; therefore, there is a risk of overlapping these signals at the receiver, which is a significant barrier for CE.

8.6 Flow Diagram

8.6.1 Input sample

The transmitter block part receives the voice input samples. The input sampling is the first block. This block takes serial data as input and provides a 2-bit IQ as an output. As a result, you'll get a collection of symbols, each

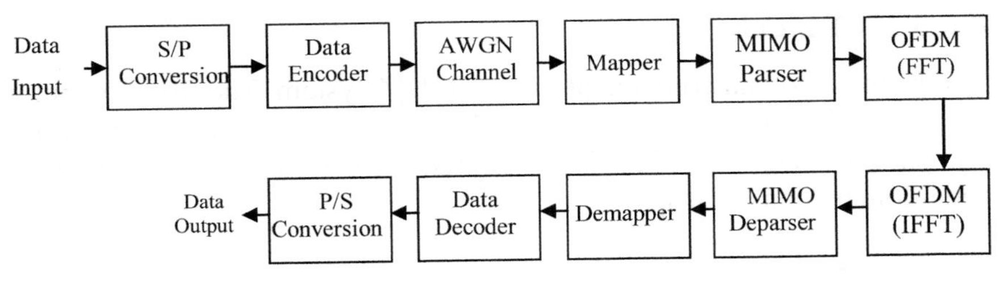

Figure 8.2 Proposed flow diagram.

with two bits. As a result, this input sampler is the component that combines two bits.

8.6.2 Serial to parallel converter

The output from the symbol mapper is the input to this block. SIPO, or serial input parallel output, is a block that turns serial data provided as input into parallel data. The serial input is sent into the seventh array, and the data is shifted to the above register per clock cycle. The data in the array is forwarded after 8 clock cycles. The SIPO output consists of eight real data registers and eight imaginary data registers.

8.6.3 AWGN channel

The AWGN noise process is a simplified noise model being used in information theory to simulate the effect of many random activities that happen in nature. It's built into the transmitter to cut down on errors.

8.6.4 Mapper

The data bits are punctured and mapped using one of the constellations stated, such as BPSK, QPSK, or QAM, depending on the modes of operation. The procedure of puncturing involves deleting some of the parity bits.

8.6.5 FFT block

This must be the most important component inside the reception area. OFDM signals are acquired by the antenna and supplied into an FFT, which converts them back to the frequency domain. The decimation in frequency (DIF)-FFT is used within this system.

8.6.6 IFFT block

It's the most crucial component of the OFDM system. The SIPO output will be the IFFT input. We'll need two IFFT modules in this system, one for real and one for imaginary. The IFFT converts frequency domain limitations into time-domain constraints.

8.6.7 BER

The BER performance of MIMO-OFDM STBC systems with various antenna configurations has been designed. Because of the larger diversity gain of these systems, the BER performance improves dramatically with more received antennae.

8.7 Module Explanation

The proposed method is divided into four module, the four modules are spitted as

- OFDM modulation/demodulation
- FFT/IFFT Block
- OFDM Transmitter
- OFDM Receiver

8.7.1 OFDM modulation/demodulation

The bit stream to be broadcast is mapped to QAM symbols, followed by a 64-point IFFT, a parallel-to-serial conversion, and a cyclic prefix of length 16 to extend the resulting time-domain sequence. Because IEEE 802.11a is a TDD technology, the FFT can be performed on the same hardware.

8.7.2 FFT/IFFT block

One radix 2 butterflies computing unit, memory blocks to cache streaming data, ROM to hold FFT twiddle factors, and control logic are all included in each FFT stage of the radix 2 FFT stage. Each stage's memory size is equal to its number.

8.7.3 OFDM transmitter

The transmitter receives data bits as inputs. These bits passed through the scrambler to disperse the input sequence and avoid the power spectrum of input signals being dependent on the actual transmitted data. The bit sequences are scrambled using Scrambler.

8.7.4 OFDM receiver

The receiver blocks are determined by the signal coding scheme employed in the transmitter. Synchronization, FFT, and MIMO detection units are the three pieces that make up the receiver. The cyclic prefix should be eliminated after getting the symbol. After that, data is sent to the FFT block.

8.8 Results and Discussion

8.8.1 Selection of voice source

The figure 8.3 Male and female recorded voice with 30-20,000 Hz is given as the input signal for processing.

The figure 8.4 The input audio waves are limited to 1000 samples and then it is processed for getting output value.

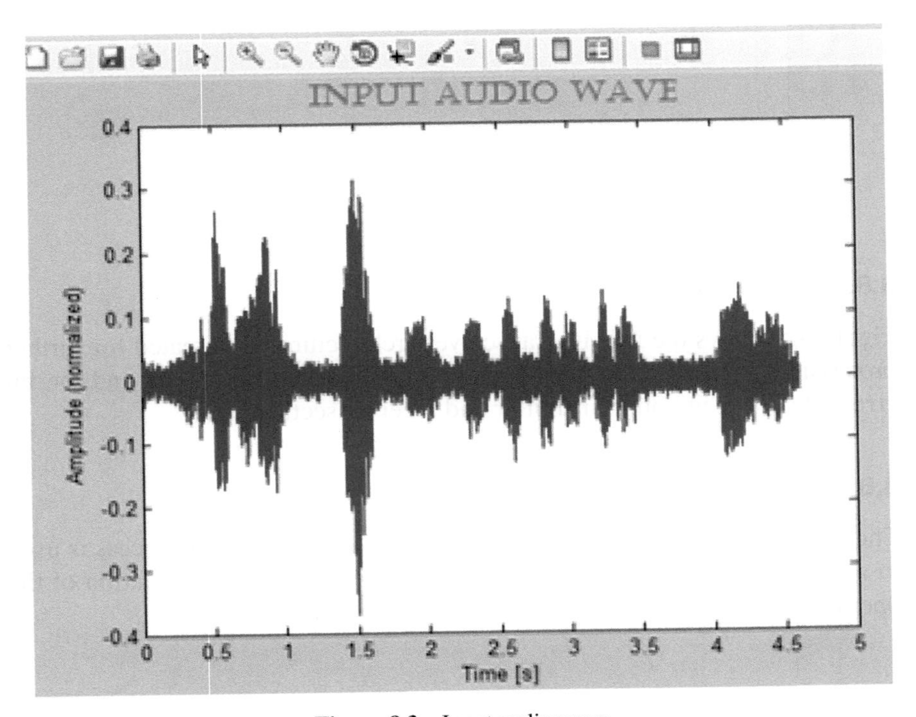

Figure 8.3 Input audio wave.

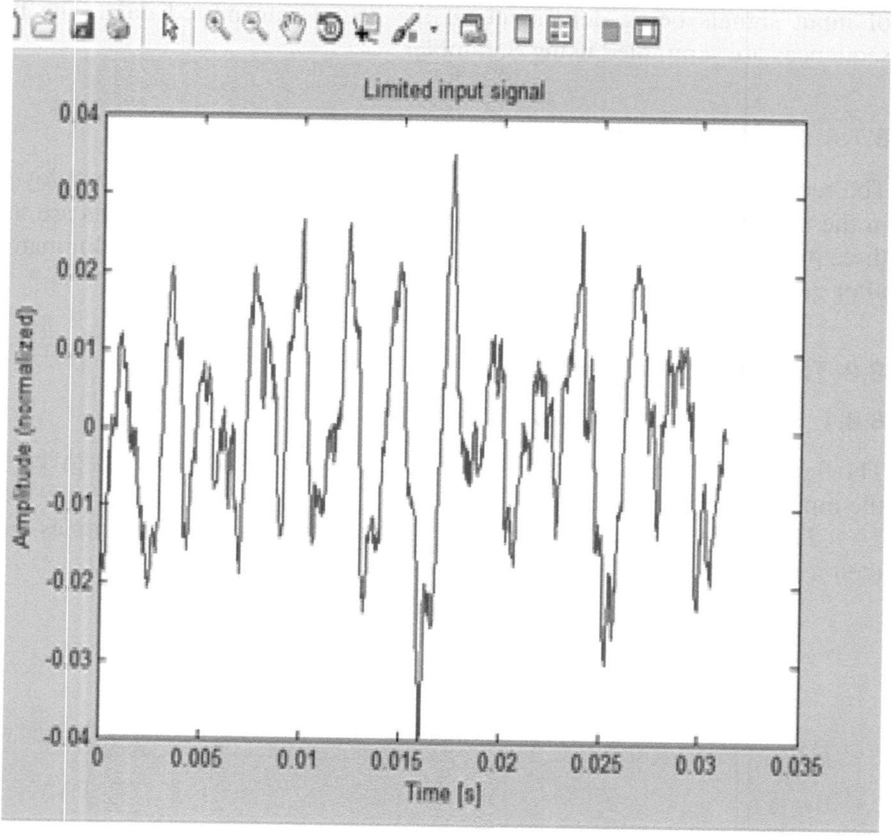

Figure 8.4 Limited input wave.

8.8.2 MIMO block design process

Figure 8.5 shows the MIMO transceiver architecture was created for further processing of input audio waves and to produce the same data and the bit error rate result in both transmitter and receiver sections.

8.8.3 Synthesis process

The Figure 8.6 the VHDL code is generated. This simulation process is used to simulate the input and output data bits. Simulation is the imitation of the operation of a real-world process or system over time.

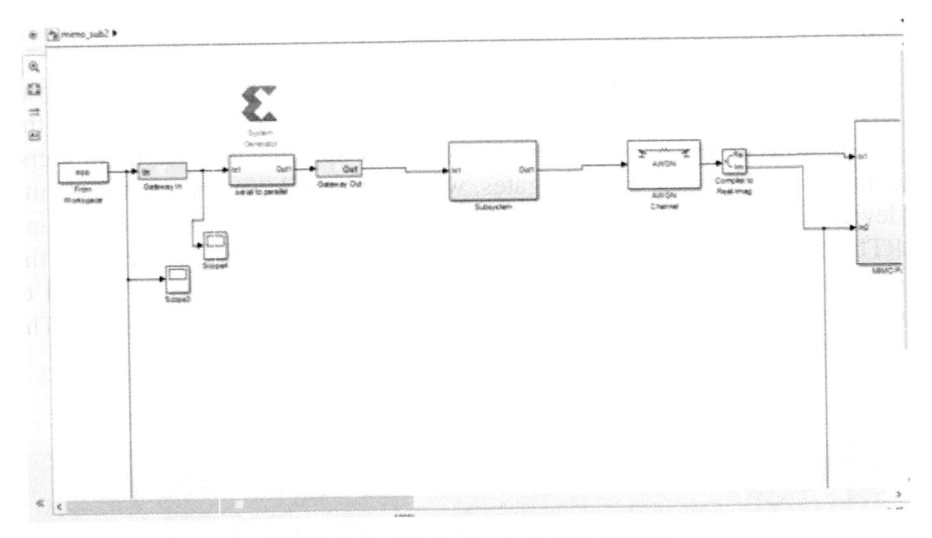

Figure 8.5 MIMO block design process.

```
2170        q => register_q_net
2171      );
2172
2173  end structural;
2174  library IEEE;
2175  use IEEE.std_logic_1164.all;
2176  use work.conv_pkg.all;
2177
2178  -- Generated from Simulink block "mimo_sub2"
2179
2180  entity mimo_sub2 is
2181    port (
2182      ce_1: in std_logic;
2183      ce_2: in std_logic;
2184      clk_1: in std_logic;
2185      clk_2: in std_logic;
2186      gateway_in: in std_logic_vector(15 downto 0);
2187      gateway_in1: in std_logic_vector(15 downto 0);
2188      gateway_in1_x0: in std_logic_vector(15 downto 0);
2189      gateway_in1_x1: in std_logic_vector(15 downto 0);
2190      gateway_in2: in std_logic_vector(15 downto 0);
2191      gateway_in3: in std_logic_vector(15 downto 0);
2192      gateway_out: out std_logic_vector(15 downto 0);
2193      gateway_out1: out std_logic_vector(15 downto 0)
2194    );
2195  end mimo_sub2;
2196
2197  architecture structural of mimo_sub2 is
```

Figure 8.6 Synthesis process.

8.8.4 RTL schematic

You can view a schematic depiction of your generated source file after the HDL synthesis portion of the synthesis process. This schematic depicts the pre-optimized design using generic symbols such as adders, multipliers, counters, AND gates, and OR gates, which are not dependent on the Xilinx device. The ISE Project Navigator is used to retrieve the VHDL codes and RTL schematics for the entire model. Inside the main model, you can see the overall RTL schematics for all of the sub-models. A ModelSim is used to create a test bench simulation to ensure that the Simulink model is valid. The RTL schematic is shown in Figure 8.7.

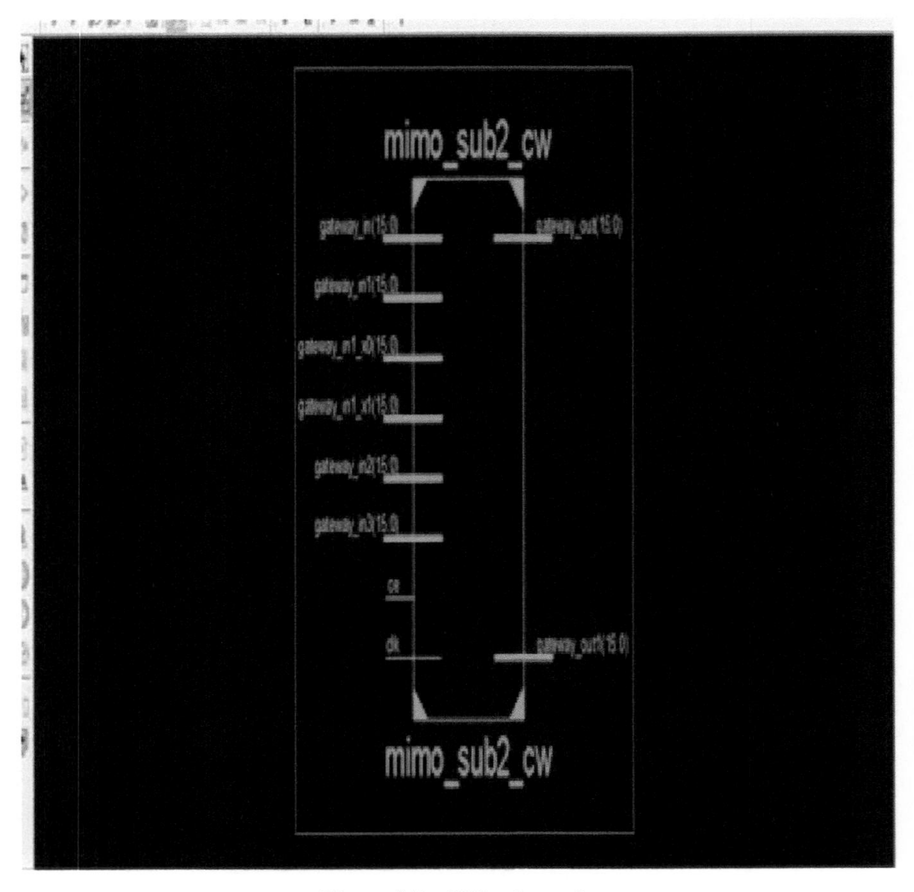

Figure 8.7 RTL schematic.

8.8.5 Technology schematic

Figure 8.8. You can view a schematic depiction of your synthesized source file after the optimization and technology targeting phase of the synthesis process. This diagram depicts the design in terms of logic pieces that are optimized for the target Xilinx device or "technology," such as LUTs, carry logic, I/O buffers, and other technology-specific components. You may view a technology-level representation of your HDL optimized for certain Xilinx architectures by looking at this diagram.

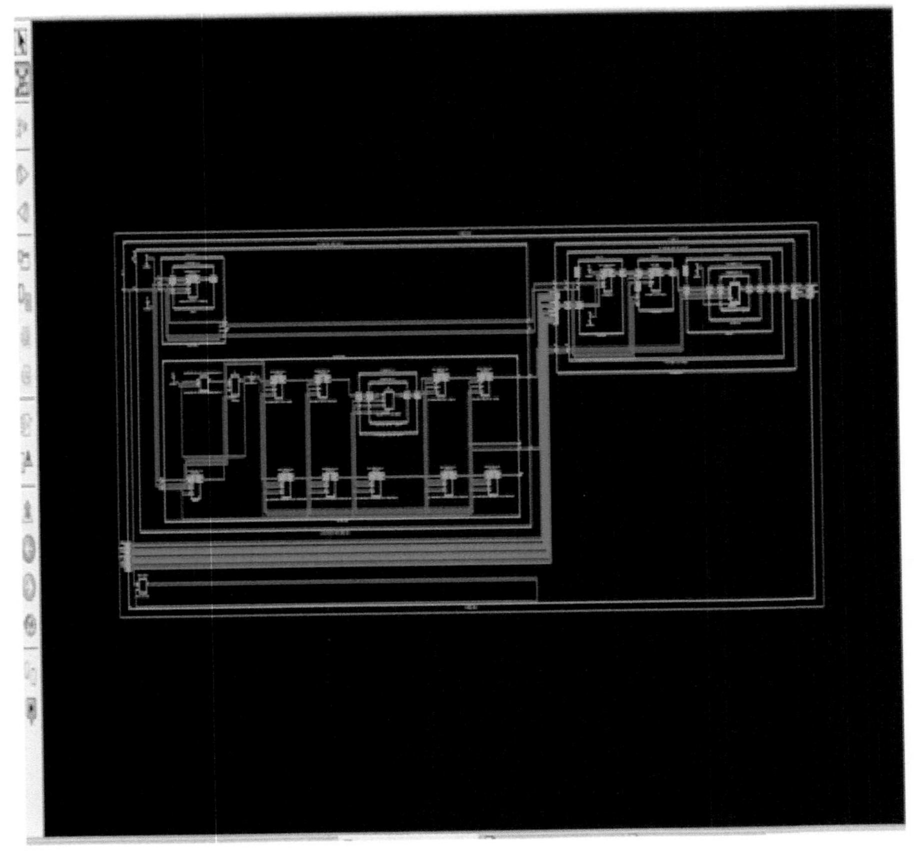

Figure 8.8 Technology schematic.

8.8.6 Power estimation

The hardware implementation of the design can bring out some important design issues like operating temperature, power consumption, time delay, and operating frequency which are very important parameters while designing a chip. The design parameter readings were taken with the help of the Xilinx System Generator.

Figure 8.9 shows the On-Chip Power panel presents the total power consumed within the device. It includes device static and user design-dependent static and dynamic power.

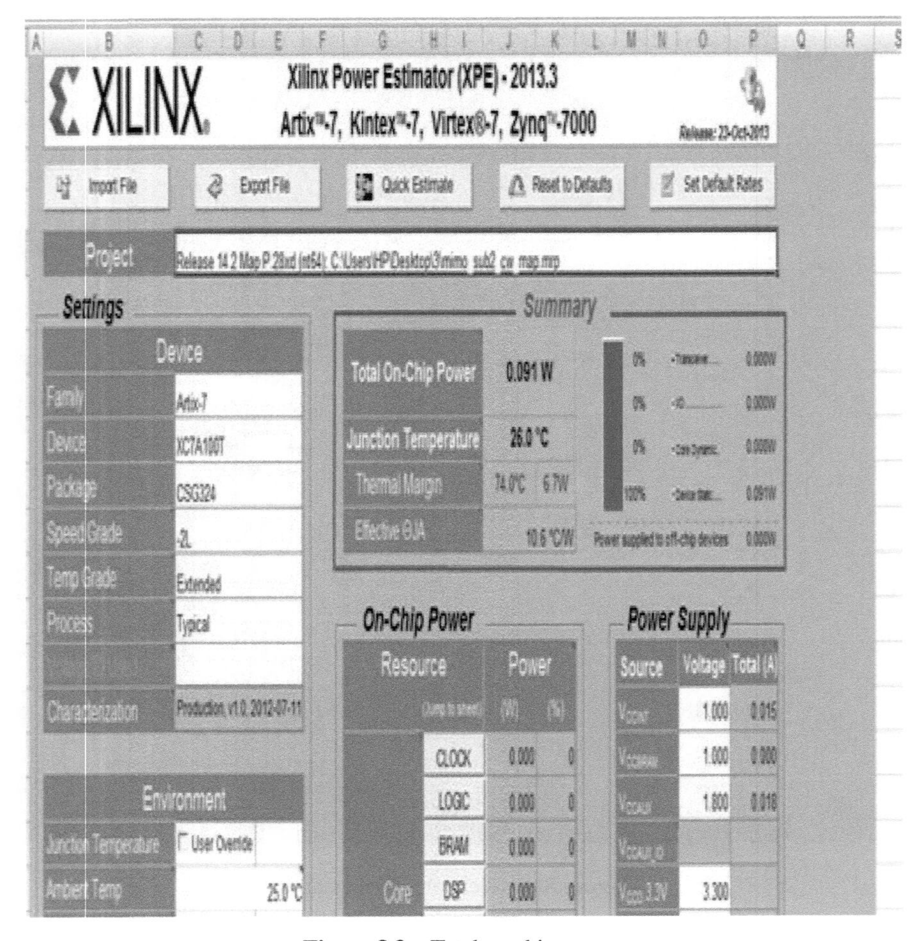

Figure 8.9 Total on chip power.

8.8.7 Static power

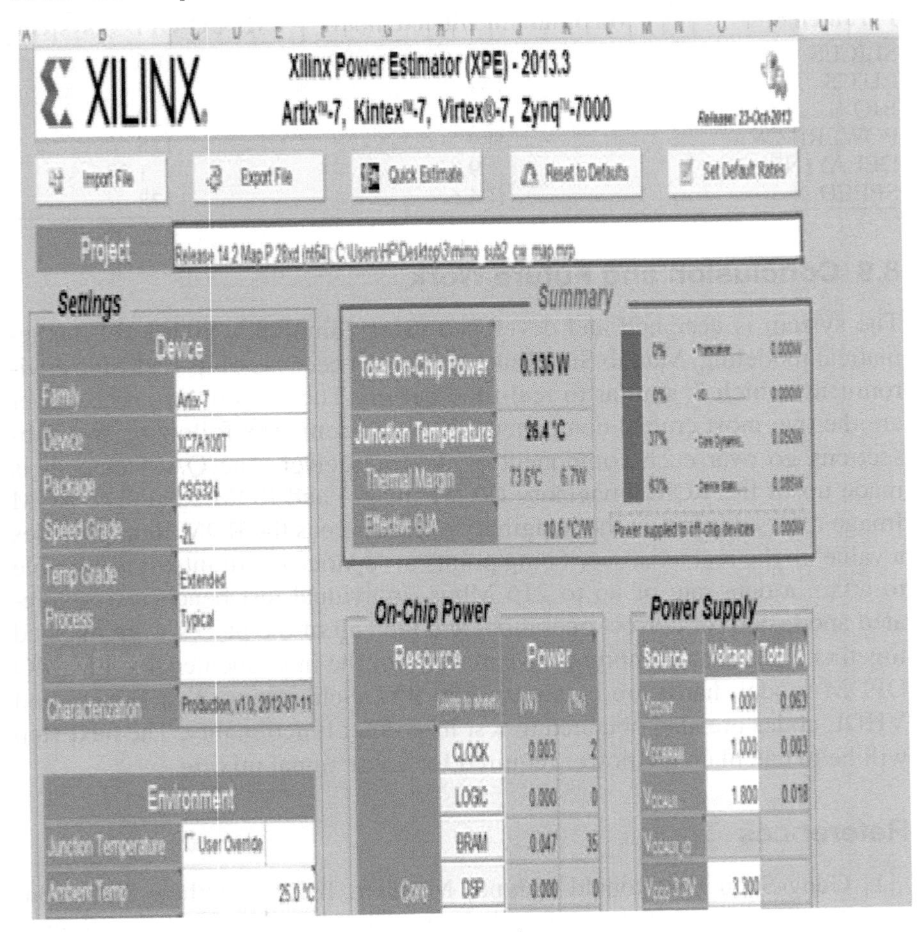

Figure 8.10 Static power.

8.8.8 Power estimation

$$
\begin{aligned}
\text{POWER} &= (\text{TOTAL ON CHIP POWER}) - (\text{STATIC POWER}) \\
&= (0.137) - (0.089) \\
&= (0.043)\text{ W (or) } 43.0\text{ mw}
\end{aligned}
$$

Table 8.1 Comparison table.

Parameters	Maximum likelihood model	MIMO-OFDM system
SLICES	158	105
LUT'S	139	105
BIO'S	184	159
POWER(MW)	589	115
DELAY(NS)	9.40	1.236
SPEED	104.6	1178

8.9 Conclusion and Future Work

The system is designed and developed using Simulink's high-level mathematical modeling. Matlab Simulink was chosen because of its real-time environment, which is similar to real-time design. The transmitter and receiver are the two most crucial components of the system. The following two subsections go over each component of the transceiver. The QAM mapper is made up of the ROM Imaginary (ROM Image) and ROM Real. The ROM Image offers a value on the imaginary axis, whereas the ROM Real provides a value on the real axis, sacrificing points in separate quadrants. It is possible to attain a data rate of up to 216 Mbps. Individual test benches were created and tested for correct operation, and then all of the pieces were mapped together to create a comprehensive model. For the implemented 4 × 4 MIMO OFDM model, hardware co-simulation, RTL Schematics, Test Bench, and VHDL codes are also obtained to test its correct functionality. The next step will be to calculate the power required for voice signal mixing.

References

[1] Geng-Shen Fu, Ronald Phlypo, Member, IEEE, Matthew Anderson, Xi-Lin Li, and Tülay Adalı, Fellow, IEEE "ICA based Blind Source Separation by Using Markovian and Invertible Filter Model "IEEE Trans on Signal Processing, vol. 62, no. 16, 2014.

[2] A. Burg, M. Wenk,M. Zellweger,M.Wegmueller, N. Felber, and W.Fichtner, "VLSI implementation of the sphere decoding algorithm," in Proc. ESSCIRC, pp. 303–306, 2004.

[3] A. Burg, M. Borgmann, M. Wenk, M. Zellweger, W. Fichtner, and H. Bolcskei, "VLSI implementation of MIMO detection using the sphere decoding algorithm," IEEE J. Solid-State Circuits, vol. 40, no. 7, pp. 1566–1577, 2005.

[4] A.Hyvärinen, "Fast and robust fixed-point algorithms for independent component analysis," IEEE Trans. Neural Netw., vol. 10, no. 3, pp. 626–634,1999.

[5] [5]Bell and T. Sejnowski, “An information maximization approach to blind separation and blind deconvolution,” Neural Comput., vol. 7, pp. 1129–1159, 1995.

[6] Belouchrani, K. Abed-Meraim, J.-F. Cardoso, and E. Moulines, “A blind source separation technique using second-order statistics,” IEEE Trans. Signal Process., vol. 45, no. 2, pp. 434–444,1997.

[7] B. A. Pearlmutter and L. C. Parra, “Maximum likelihood blind source separation: A context-sensitive generalization of ICA,” in Advances in Neural Information Processing Syst. 9. Cambridge, MA, USA: MIT Press, pp. 613–619, 1997.

[8] C.-J. Huang, C.-W. Yu, and H.-P. Ma, “A power-efficient configurable low-complexity MIMO detector,” IEEE Trans. Circuits Syst. I, vol. 56, no. 2, pp. 485–496, 2009

[9] G. Knagge, M. Bickerstaff, B. Ninness, S. R. Weller, and G. Woodward,“A VLSI 8 8 MIMO near-ML decoder engine,” in Proc. IEEE SIPS, pp. 387–392,2006

[10] H. Attias, “Independent factor analysis with temporally structured sources,” in Proc. Adv. Neural Inf. Process. Syst., pp. 386–392, 1999

[11] H. Snoussi and A. Mohammad-Djafari, “Bayesian unsupervised learning for source separation with mixture of Gaussians prior,” J. VLSI Signal Proces. Syst. Signal, Image, Video Technol., vol. 37, no. 2-3, pp. 263–279, 2004.

[12] K. waiWong, C. ying Tsui,R. S.-K. Cheng, and W.hoMow, “AVLSI architecture of a K-best lattice decoding algorithm for MIMO channels,” Proc. ISCAS, vol. 3, pp. 273–276, 2002.

[13] M. Mahdavi and M. Shabany, “Novel MIMO detection algorithm for high-order constellations in the complex domain,” IEEE Trans. VLSI Syst., vol. 21, no. 5, pp. 834–847,2013.

[14] R. Guidara, S. Hosseini, and Y. Deville, “Blind separation of nonstationary Markovian sources using an equivariant Newton-Raphson algorithm,” IEEE Signal Process. Lett., vol. 16, no. 5, pp. 426–429, 2009.

[15] S. Hosseini, C. Jutten, and D. T. Pham, “Markovian source separation,” IEEE Trans. Signal Process., vol. 51, no. 12, pp. 3009–3019, 2003.

[16] X.-L Li and T. Adalı, “Independent component analysis by entropy bound minimization,” IEEE Trans. Signal Process., vol. 58, no. 10, pp. 5151–5164, 2010.

[17] X. Chen,G.He, and J.Ma, “VLSI implementation of a high-throughput iterative fixed-complexity sphere decoder,” IEEE Trans. Circuits Syst. II, vol. 60, no. 5, pp. 272–276, 2013.

[18] Yeredor,"Blind separation of Gaussian sources via second-order statistics with asymptotically optimal weighting," IEEE Signal Process. Lett., vol. 7, no. 7, pp. 197–200, 2000.

[19] Z. Koldovský, P. Tichavský, and E. Oja, "Efficient variant of algorithm fastica for independent component analysis attaining the Cramér-Rao lower bound," IEEE Trans. Neural Netw., vol. 17, no. 5, pp. 1265–1277, 2006.

[20] Z. Guo and P. Nilsson, "Algorithm and implementation of the K-best sphere decoding for MIMO detection," IEEE J. Sel. Areas Commun., vol. 24, no. 3, pp. 491–503, 2006.

Chapter 9

Empowering Radio Resource Allocation to Multicast Transmission System Using Low Complexity Algorithm in OFDM System

R. Kabilan[1], R. Ravi[2], J. Zahariya Gabriel[3], and R. Mallika Pandeeswari[4]

[1]Associate Professor, Department of ECE, Francis Xavier Engineering College, Tirunelveli, India
[2]Professor, Department of CSE, Francis Xavier Engineering College, Tirunelveli, India
[3]Associate Professor, Department of ECE, Francis Xavier Engineering College, Tirunelveli, India
[4]Full Time Ph.D Scholar, Dept of ECE, Francis Xavier Engineering College, Tirunelveli, India
Email: rkabilan13@gmail.com; fxhodcse@gmail.com; zahagabs@gmail.com; mallikapandeeswari123@gmail.com

Abstract

In an OFDM system, radio resources are allocated and managed to ensure optimum channel quality for customers. Using a subgroup approach, subscribers have been categorized into different groups based on channel quality. The resources allocated to a subgroup's consumers are power allocation, capacity, and data rate. A simple algorithm The OFDMA multicast method, Frequency domAIN Subgroup algoriThm (FAST), is required to reduce the dimensionality of either subgroup creation problem and also the performance restrictions of basic multiplex data delivery strategies. FAST increases the number of allowable subgroups as well as seeks the best subgroup that outperforms the previous iteration. Throughput increases as a result of this approach due to increased system capacity. The simulation results demonstrate that such a system achieves near-optimal achievement with such a restricted computational load, and therefore, the greater the spectral efficiency, the more resources are assigned.

9.1 Introduction

Wireless communication refers to the transmission of data among two or more sites that are not connected by an electrical conductor. A large percentage of wireless technologies make use of radio. Radio waves travel short distances, like a few meters in television, as well as thousands, though not millions, of kilometers throughout deep-space radio communications. Two-way radios, cellular phones, PDAs, as well as wireless networking are examples of fixed, mobile, as well as portable applications. Wireless effectively approaches services such as long-distance communications that would have been impossible to implement through cables [3-6]. The term is frequently used in the telecommunications industry to refer to information-carrying systems that do not rely on wires but instead rely on some form of energy.

The use of multiple antennas on the receiving side of a wireless link is referred to as receive diversity [9]. This also contributes to job completion on time and increases productivity. The impacted regions could indeed receive support and assistance from the use of these updates as well as timely wireless communication [11]. Different users can be assigned a different number of sub-carriers to provide distinct QOS, allowing every user's data rate as well as error likelihood to be controlled independently [10].

OFDMA can be considered an alternative to communicating using OFDM and TDMA. Low-data-rate allows users to send continuously at low transmission power without using a continuous wave high-power carrier. It's indeed possible to develop both a constant and a shortened delay [11–13]. OFDMA is supposed to be a key fit for broadband wireless networks due to its scalability, MIMO-friendliness, and potential benefit of channel frequency selectivity.

Coherent adaptation to the new and yet coherent detection and decoding strategies is required for random signal configurations [2]. One-dimensional connection predictions are frequently used within OFDM systems to obtain an exchange between complexity and accuracy. The first method, block-type pilot channel estimation, is based on an assumption of a slow fading channel as well as involves inserting pilot tones into all OFDM symbol subcarriers over a predetermined time period [14].

9.2 Existing System

Greater organization implementations, including multimedia uploading, mobile TV, and others, are becoming increasingly important in today's climate of quick online service growth. Developing the right multicast drive techniques in OFDMA-based systems has also proven to be a difficult task, prompting numerous research efforts.

9.2.1 Conventional multicast scheme

CMS cells all receive the very same data rate. It broadens the system's coverage. However, as mentioned in the preceding member, the MCS is chosen with the lowest channel quality. As a result, the OFDMA strategy is not fully utilized as the sample sizes grow.

9.2.2 Radio resource management (RRM) algorithm

The base station allocates various resources to the subgroup's customers based just on the CSI provided by the multicast on each scheduling frame. Also, every user is assigned an MCS, as well as a CSI with the highest MCS chosen by the base station. The RRM chooses the appropriate subgroup formation technique based on the CSI data.

9.3 Proposed System

To overcome the throughput shortcomings of conventional multicast data transmission lines, a subgrouping technology is used that divides users into different groups based just on the channel quality encountered. The frequency-domain subgroup algoriThm (FAST) is being used to reduce computing complexity by dividing multicast receivers into the different subgroups and allocating transmission resources to them based on experienced channel conditions.

9.4 Multi Rate Scheme

Classifies subscribers into subgroups based on the quality of the channel.

Chosen MCS and allocate resources to every enabled sub-group, resulting in greater processing capacity.

Figure 9.1 illustrates the elements of such a multiplex subgrouping system

9.4.1 OFDMA framework

The primary concept underlying OFDMA systems is the division of the obtainable frequency spectrum into and out of multiple subcarriers. The

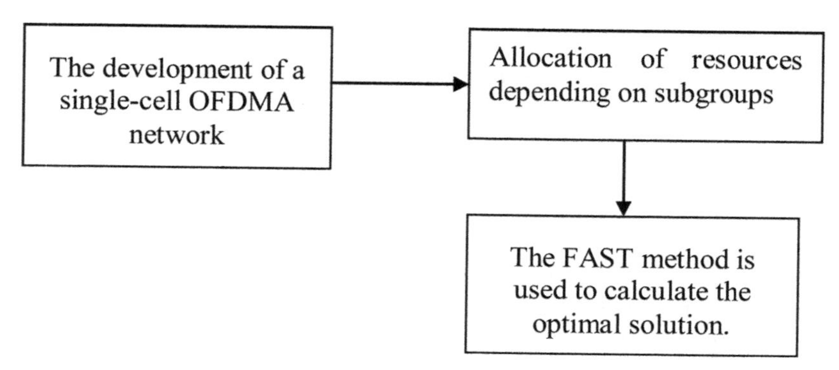

Figure 9.1 Blocks involved in multicast sub grouping system.

frequency components of both the subcarriers have been overlapped as well as orthogonal to provide high spectral efficiency, hence the name OFDMA. The binary information is retrieved after demodulation and channel decoding.

9.4.2 Utilisation of resources depending upon subgroups

Subscriptions are divided into subgroups using the subgroup approach. Data rates are assigned to consumers for each category depending on the power allocation as well as channel quality. The CSI response is predicated on the reasoned SINR as well as being a measure of the channel quality of a terminal. Users in the highest MCS subgroup will have better channel quality, resulting in a higher data rate as well as throughput.

9.4.3 Channel state information (CSI)

In wireless communication, channel state information is linked to the known channel characteristics of a communication link. This data depicts the cumulative effect of dispersion, fading, as well as power decay with distance as a signal travels from the transmitter to receiver.

9.4.4 Signal to interference plus noise ratio (SINR)

It is a number used to compute the maximum theoretical limits for wireless systems, such as networks. In contrast, zero interference diminishes this same SINR to SNR, which again is less commonly used in statistical equations of wireless networks such as cellular networks.

9.5 Frequency Domain Subgroup Algorithm (FAST)

The Frequency domain subgroup algorithm (FAST) performs multicast subgrouping successfully and at a low processing cost. Depending on the CSI response from the base station, FAST determines the best subgroup set-up for each schedule frame. The FAST behavior is a low-cost feature for forming subgroups.

9.6 Results and Discussions

Subscriptions are subclassified depending on the input quality using a subgroup technique. Power allocation, capacity, as well as data rate are among the resources assigned to service users in the subgroup. To reduce the computational complexity of the subgroup, a low-complexity algorithm termed Frequency DomAin Subgroup algoriThm (FAST) for OFDMA multipathing systems is used. The simulation results show that this system achieves near-optimal performance at a low computational cost and that the greater the spectral efficacy, the more and more resources are allocated.

9.6.1 Separate cell's creation

Figure 9.2 depicts the creation of every cell inside the multicast subgrouping technique, which would be symmetrical to prevent signal leakage.

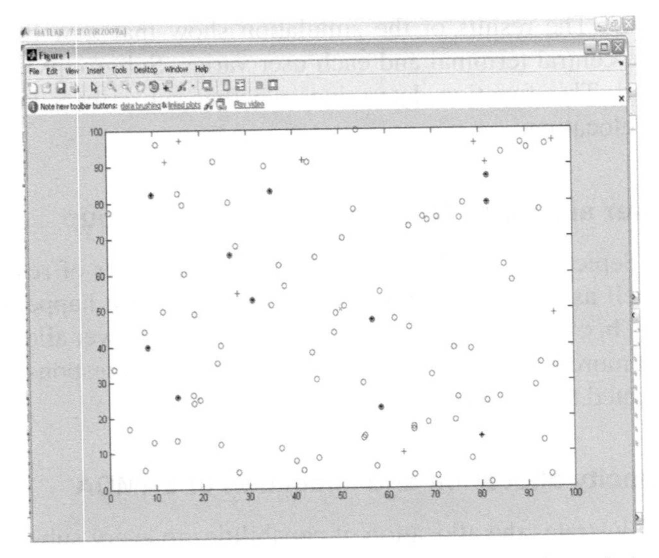

Figure 9.2 Creation of each cell in multicast sub grouping technique.

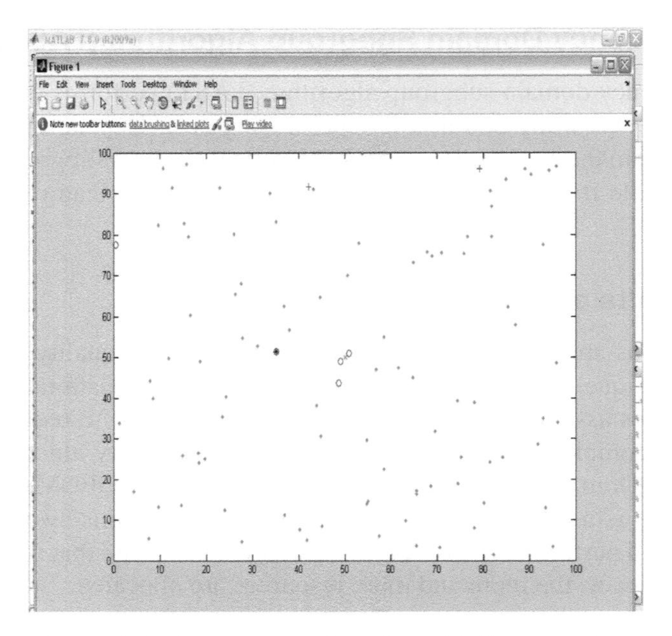

Figure 9.3 Average energy in each user Vs Round number.

9.6.2 After every round, the average energy in each user

For each round number, the average energy within every node can be seen in Figure 9.3. The results of the simulation show that now the separation between the central terminal and each user varies. Subgroups of users have been formed. This function determines whether subscribers in a subgroup have been relocating.

9.6.3 Power allocation to each and every subgroup

Figure 9.4 depicts the number of subgroups, the number of resources allocated, as well as the predicted as well as channel noise happening in each subgroup (a, b, c). If the expected noise decreases, the power allocation might be assigned more efficiently. If the expected subchannel response is poor, the power allocated to that subchannel is reduced as well.

9.6.4 Capacity allocation and allocation of LAMDA

Figure 9.5 illustrates the allocation of capabilities to every subgroup for just the provided data rate and verifies that it is correct. The capacity but instead

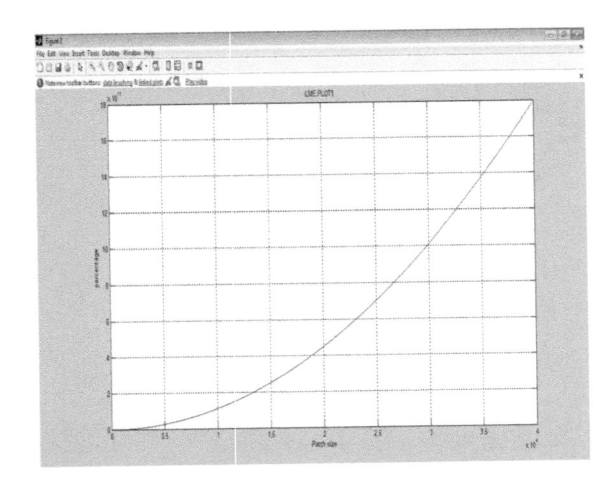

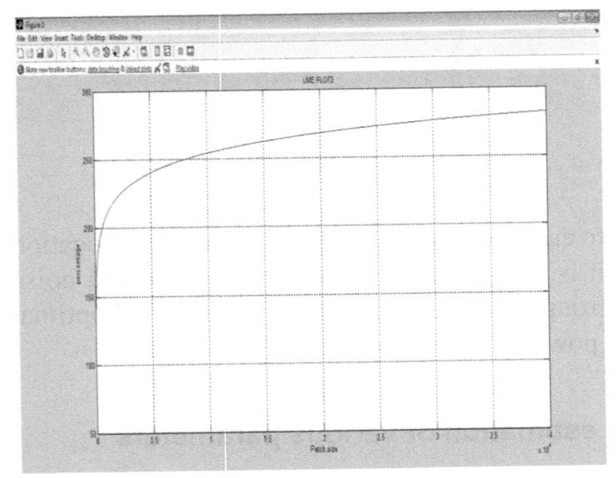

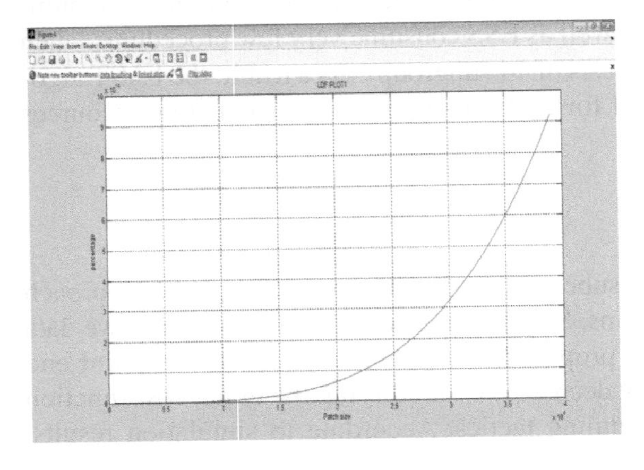

Figure 9.4 (a,b,c) Amount of resources allocated.

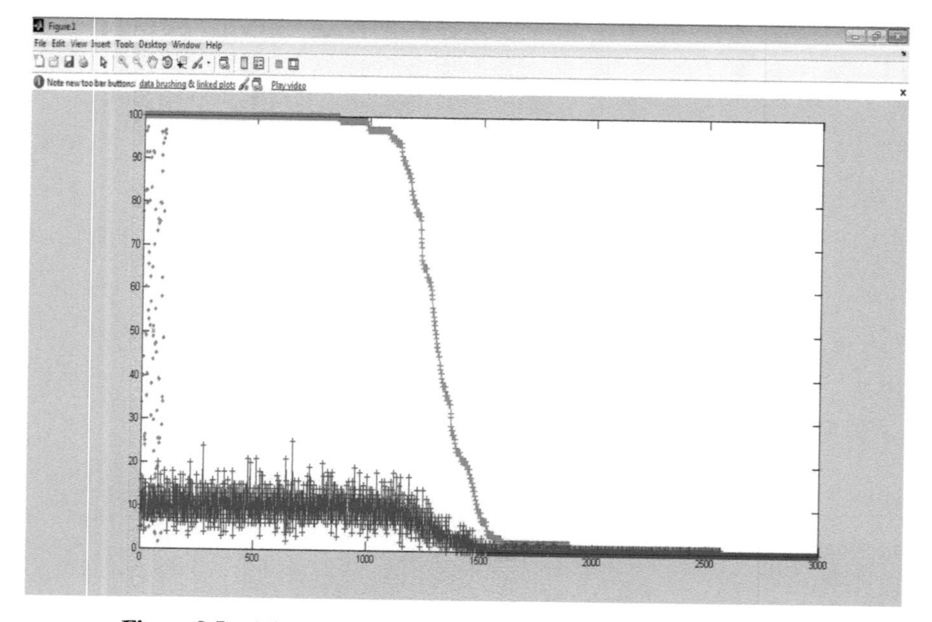

Figure 9.5 Allocation of capacity and wavelength for each subgroup.

wavelength is then assigned to each user inside a subgroup, and the allocation has been checked to see if it is correct or not based on the expected noise within every user in the subgroup. The optimization values are ECSI optimal as well as ECSI comparable power.

9.6.5 The performances estimation of various parameters

Figure 9.6 depicts the number of operations necessary in relation to the number of subgroups. MGGA as well as ESS consume very few procedures when the output configuration consists of 14 subgroups. As a result, the FAST algorithm's performance is ideal for partitioning users and allocating resources to them.

9.7 Conclusion

FAST is a low-complexity subcategory distribution of resources approach for OFDMA multicast systems. Standard multicast methodologies have data transfer limitations, but the proposed policy overcomes this constraint and ensures system functionality decreases by modifying the target cost function to handle a variety of scheduling tactics. According to simulation results,

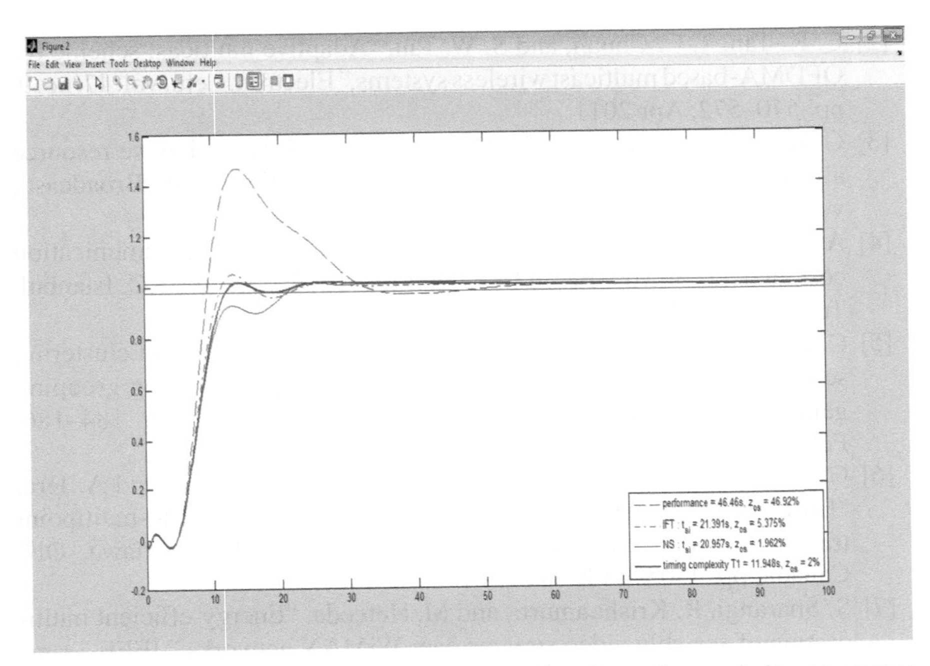

Figure 9.6 Comparison of various parameters vs number of operations needed to convergence.

the proposed algorithm method effectively reduces each subgroup formation, helps to ensure improved performance going forward towards the precise detection method both in the overall approach as well as roughly equivalent equality dispersion, and needs fewer iterations to accomplish convergence than existing methods.

A resource allocation method based on the dual optimization tool will be considered in the future to maximize OFDMA system throughput while meeting the QOS criteria of both real-time as well as best-effort traffic more than a time-varying channel. The other one is the tolerable average absolute variance of the rate of transmission for RT services, which is then used to manage transmission rate fluctuations as well as minimize RT packet delay.

References

[1] Araniti, Giuseppe; Condoluci, Massimo; Iera, Antonio; Molinaro, Antonella; Cosmas, John; Behjati, Mohammadreza. A Low-Complexity Resource Allocation Algorithm for Multicast Service Delivery in OFDMA Networks. IEEE Transactions on Broadcasting, 60(2), 358–369. 2014.

[2] C. K. Tan, T. C. Chuah, and S. W. Tan, “Adaptive multicast scheme for OFDMA-based multicast wireless systems,” Electron. Lett., vol. 47, no. 9, pp. 570–572, Apr 2011.

[3] G. Araniti, M. Condoluci, L. Militano, and A. Iera, “Adaptive resource allocation to multicast services in LTE systems,” IEEE Trans. Broadcast., vol. 59, no. 4, pp. 658–664, Dec 2013.

[4] A. Alexious, C. Bouras, V. Kokkinos, and G. Tsichritzis, “Communication cost analysis of MBSFN in LTE,” in Proc. IEEE 21st PIMRC, Istanbul, Turkey, pp. 1366–1371, Sep 2010.

[5] C. K. Tan, T. C. Chuah, S. W. Tan, and M. L. Sim, “Efficient clustering scheme for OFDMA-based multicast wireless systems using grouping genetic algorithm,” Electron. Lett., vol. 48, no. 3, pp. 184–186, Feb. 2012.

[6] G. Araniti, V. Scordamaglia, M. Condoluci, A. Molinaro, and A. Iera, “Efficient frequency domain packet scheduler for point-to-multipoint transmissions in LTE networks,” in Proc. IEEE ICC, Ottawa, ON, Canada, pp. 4405–4409, Jun. 2012.

[7] S. Sharangi, R. Krishnamurti, and M. Hefeeda, “Energy-efficient multicasting of scalable video streams over WiMAX networks,” IEEE Trans. Multimedia, vol. 13, no. 1, pp. 102–115, Feb 2011.

[8] R. Giuliano and F. Mazzenga, “Exponential effective SINR approximations for OFDM/OFDMA-based cellular system planning,” IEEE Trans. Wireless Commun., vol. 8, no. 9, pp. 4434–4439, Sep 2009.

[9] J. F. Monserrat, J. Calabuig, A. Fernandez-Aguilella, and D. Gomez-Barquero, “Joint delivery of unicast and E-MBMS services in LTE networks,” IEEE Trans. Broadcast., vol. 58, no. 2, pp. 157–167, Jun 2012.

[10] T. Jiang, W. Xiang, H. H. Chen, and Q. Ni, “Multicast broadcast services support in OFDMA-based WiMAX systems,” IEEE Commun. Mag., vol. 45, no. 8, pp. 78–86, Aug 2007.

[11] A. Richard, A. Dadlani, and K. Kim, “Multicast scheduling and resource allocation algorithms for OFDMA-based systems: A survey,” IEEE Commun. Surv. Tutor., vol. 15, no. 1, pp. 240–254, Feb 2013.

[12] M. Condoluci, L. Militano, G. Araniti, A. Molinaro, and A. Iera, “Multicasting in LTE-A networks enhanced by device-to-device communications,” in Proc. IEEE GLOBECOM, pp. 573–578, Dec. 2013.

[13] P. K. Gopala and H. E. Gamal, “Opportunistic multicasting,” in Proc. 38th Asilomar Conf. Signals, Syst. Comput., pp. 845–849, Nov 2004,

[14] T. P. Low, M. O. Pun, Y. W. P. Hong, and C. C. J. Kuo, “Optimized opportunistic multicast scheduling (OMS) over wireless cellular networks,” IEEE Trans. Wireless Commun., vol. 9, no. 2, pp. 791–801, Sep 2009.

[15] L. Zhang, Z. He, K. Niu, B. Zhang, and P. Skov, "Optimization of coverage and throughput in single-cell E-MBMS," in Proc. IEEE 70th VTC, Anchorage, AK, USA, pp. 1–5, Sep 2009.

[16] Y. Jiao, M. Ma, Q. Yu, K. Yi, and Y. Ma, "Quality of service provisioning in worldwide interoperability for microwave access networks based on cooperative game theory," IET Commun., vol. 5, no. 3, pp. 284–295, Feb 2011.

[17] J. Y. L. Boudec, "Rate adaptation, congestion control, and fairness: A tutorial," Ecole Polytech. Fed. Lausanne, Lausanne, Switzerland, Feb 2005.

[18] S. Deb, S. Jaiswal, and K. Nagaraj, "Real-time video multicast in WiMAX networks," in Proc. IEEE 27th INFOCOM, Phoenix, AZ, USA, pp. 1579–1587, Apr. 2008.

[19] A. Alexious, C. Bouras, V. Kokkinos, A. Papazois, and G. Tsichritzis, "Spectral efficiency performance of MBSFN-enabled LTE networks," in Proc. IEEE 6th WiMob, Niagara Falls, ON, Canada, pp. 361–367, Oct. 2010.

[20] C.Mehlfuhrer et al.,"Simulating the long term evolution physical layer", in Proc. 17th EUSIPCO, Zadar, Croatia, pp.1471-1478, Aug.2009.

Chapter 10

Survey on RF Coils for MRI Diagnosis System

K. Sakthisudhan[1], N. Saranraj[2], and C. Ezhilazhagan[3]

[1]Professor, Dr. N.G.P. Institute of Technology, Coimbatore, Tamilnadu, India.
[2]Research Scholar, Anna University, Chennai, Tamilnadu, India.
[3]Assistant Professor, Dr. N.G.P. Institute of Technology, Coimbatore, Tamilnadu, India.
Email: drkssece@gmail.com; saran.plc.1963@gmail.com; ezhisang20@gmail.com

Abstract

An MRI coil provides a flexible and compact design for medical diagnosis application and perfectly suits human anatomy structure. It is enhanced for a variety of radiology applications. These coils are made up of superconducting materials with high temperature, analyze inverse methodology through a huge dynamic view of body centric application, 3D parallel diagnosis analysis with larger channels. It offers sufficient Signal to Noise Ratio (SNR), excited homogeneity and sensitivity and high resonant frequency, and Q-factor. In this paper, research offers the various classes of radiology methods with scientific, industrial, and medical benefits. Furthermore, the research work offers the current trends and development of radiology techniques. The drawbacks of various MRI coils and the development of specific applications were also discussed.

10.1 Introduction

A huge proportion of diagnostic and research imaging methods are using Magnet Resonance Imaging (MRI). It is the most distinguished method in its capacity for going beyond the anatomy imaging as it provides the reverse technique like Ultrasound-based non-invasive therapy, nearest neighbor conventional non accelerated cardiac 3D parallel imaging, and advanced measuring methods [1]. It also offers an excellent contrast to high space and time resolutions. MRI offers the most advanced methodologies for measuring intraocular resolutions. The MRI hardware is substantial, such as a homogenous distribution of the magnetic field and the flip angle with a moderate gradient. Advances in MRI technology are to ensure more relaxed patients that need scans while offering higher images of higher spatial and temporal resolution, high SNR, restricted Specific Absorption Ratio (SAR), high-quality factors, and a shortened period of MRI results. For an MRI test, MRI coils are important elements as they are responsible for the MR signal elation and reception. In terms of therapeutic and MR technologists, this survey on MRI Coils and its basic tenets is divided into three parts. MRI coils are illustrated with the basic concepts and phraseology of the application fields. Then, the latest state-of-the-art transmission and reception of spools are thoroughly explained. The last section addresses advanced technology, and techniques with the current growth in MRI Coils. The entire paper contains topic-oriented sources and supplementary recommendations for literature.

A modern micro coil design [2] provides the orientation of both unit and tip position in size and layout comparable with conventional ones, and the nearest micro stream (Radio Frequency) RF coil has the capacity to design [3] a vast array of typical surface bobbins at high frequencies without RF. To illustrate the principle in the RF bobbins, [4] present a 400-MHz micro-strip transmission line (MTL) RF bow with the second harmonic resonance on 9.4 tesla in the MTL resonator for rat imagery and then [5] equate it to a receiver or transceiver transmitting efficiency by using a single eight-element bowl array. In optimizing these coils for sensitivity encoding (SENSE) activity, it has included the effects of physiologic noise. [6] in the way to instruct their RF spectrum sensitivity of flexible components, electrical and magnetic behavior specifications, biological phantom model was performed, and suggestion of sufficient numbers of coils was included for achieving better SNR than homogeneous coils, which are located on anterior and posterior sides of the body. Additionally, [7] 128 numbers of channels in the MRI system with a receiver in the body system consist of both 2D and large-volume acquisitions for highly accelerated body imaging. [8] An actively decoupling saddle and surface spiral environment achieves a homogeneous excitation with high sensitivity for nuclei with a close resonance frequency (19F) MRI while holding enough nuclei with close resonance frequency (1H) signal. SNR for anatomical images [9], high quality and high-spatial 2D Cine-angiography (CINE) cardiovascular picture at 7.0 with parallel pictures and [10] to the multiple-element surface spiral, and no longer restricted by a single picture. A novel concept for a monopoly array pattern is to overcome dipole design challenges [11], in which offers the dipole antenna in the brain imaging of human beings is deterred by a long antenna length and the design and efficiency of a 31 channel Positron Emission Tomography (PET) and Magnetic Resonance (MR) brain array coil, simultaneous discussions are done in the following sections.

10.2 Survey of Literature

10.2.1 Design of transceiver RF coils

The SNR in a Nuclear Magnetic Resonance (NMR) experiment is determined by several factors. Two more successful experiments emerged in the early years of MRI: the first is to make a better gradient and pulsating design sequence in order to improve spatial resolution and reduce the imaging time, and the second was to improve the technology used for RF coil. Irrespective of field power, a recipient coil's principal requirement is to attain maximum SNR to get the best quality of image. The SNR value fundamentally reflects the trust we have in demographics strength in characteristics of imaging (or MR spectroscopy). The physical features in the sample image are determined in an image as a visible feature (bright spot at the given location) when the signal is strong enough then the intensity of random noise variations. The noise at the initial stage is the Gaussian white noise when the recipient is detected but is generally taken in Rician form and further affected by the process of reconstruction, in particular with Array Coils[12, 13].

Each physical experiment comprises random or systemic noise that can severely affect measurement accuracy. It includes the extent to which the noise influences an experiment usually by the SNR. Random signal, superposed on the true signal, can be considered MRI noise. Because of the random character of the noise level, the mean value is zero, and thus the normal noise deviation is the objective measure of the noise level. The SNR is improved by increasing the number of scans i.e. repeating, called ‘signal averaging’. The factor of two SNR enhancements can be used by RF coils to reduce scanning time by a factor of 2. The signal average is compromising the extra scanning time required for obtaining the data [14]. The signal for each MRI experiment originates from the rotating dipole in the transverse plane [15]. Two properly aligned RF spools can catch both dimensions of the rotating field while a single recipient spindle can take up a single dimension from this rotating field in Linear Polarized (LP) mode. It consists of the Circular

Polarized (CP) model of the coil pair using a Hybrid Square which summarizes the two signals on both coils with a 90^0-phase difference [16]. This is an E-MRI signal polarization concept (e-MRI) similar to optics, since polarized lenses only allow some restricted light beams to penetrate along the glass, which is also the way that they block all of the other polarized radii [17]. In the case of MRI, this is a signal polarization concept that is comparable. The benefit of the CP model in TX spools is that the linearly polarized device needs half its power to supply the same B1 + field.

10.2.2 The development of RF based MRI coils

The existing method uses a vivo proton or NMR core applications of 7T with the latest form of high-frequency RF coil. The method is used to create a collection of echo images with a gradient recalled by single-turn and two-turn RF surface bowls from both the picture and human intelligence, which are caused by a reduced loss of radiation and lower output of the RF sample load compared to traditional surface coils. This method is used to create a set of gradient-recognized echo images. The coil is defined by a high Q factor, with no RF shielding, the smaller size of the coil, reduced cost, and simple manufacture [3] and describes a new single microwave structure with three different winding components, both providing orientation to the system and the tip location. This makes the architecture of the micro belt ideal for much versatile interventions of devices. A 0.2T MRI system having moderate gradient efficiency has been demonized in real-time reliably tracking the three points of an intervention device, which results in an important inclusion to an active MRI system where the tracking methods provide both device orientation and size on the tip of location information and the configuration [2]. The reverse technique defined in [18] supports the conception of RadioFrequency (RF) coils for MRI applications using two techniques as follows. The first method is to measure current on spin and shield cylinders, which produces a certain internal magnetic field using the time-harmonic electric Green function. Second technique, streaming function moment is used to implement the theoretical current density in an RF spiral, resulting in that the current being measured in the cylinder which produces a particular magnetic field, and Green's function calculates currents in a cylinder coil and a cylinder of the shield and [19] uses an improved model for designing the RF MRI High-Temperature Supreme (HTS) conductor coils. The RF HTS product has a coildiameter of 65 mm for 0.2T and 1.5T MRI systems measuring 19K and 23K respectively. The RF HTS coil is designed and manufactured to analyze the resonant frequency and spinal frequency accurately. The ballon injured procedure for inherited hyperlipidemia in rabbits was treated by Watanabe introducing an experimental model of atherosclerosis by an Intra-Vascular (IV) MRI probe for high resolution atherosclerotic-ic in imaging. The new IV MRI sensor is 1.3 mm in diameter and operated by a guide wire; the effect is that both the external phase array coil and the IV MRI coil have been added to MRI and histopathological findings are associated in [20].

Introducing the development of parallel MRI technique in and new applications for interactive MRI in real-time in which emphasizes the need to evaluate the performance achieved through the increase in the ability of MRI phased array systems from standard IV to standard VIIIforhigh bandwidth channels. The system provides a forum to assess the many real-time RMI channel applications and understand the factors optimizing the choice of array size. Vaughan et al. have created an effective new body coil, Transversal Electro-Magnetic (TEM) field, and demonstrated its application of it to human studies at field forces up to 4T. We have modeled five unloaded bobbins for a finite time divergence domain and the result is obtained with a body coil. [21].

The high-frequency RF winding belt is designed using high and very high fields; considering the traditional RF winding design difficulties and limitations were introduced by [4]. This technique produces 400 MHz RF winding successfully used for rat imaging coil having the second harmony at a micro-sectional resonant field. Newly designed 32 element receiver coils is be arranged for cardiac imaging as shown in figure 10.1, consisting of 21 copper rings in the anterior range with 75mm diameter and 11 copper rings in the rear range of 107mm diameter arranged in the hexagonal lattice structure by [22]. This method is used in the processing of micro-mounting,

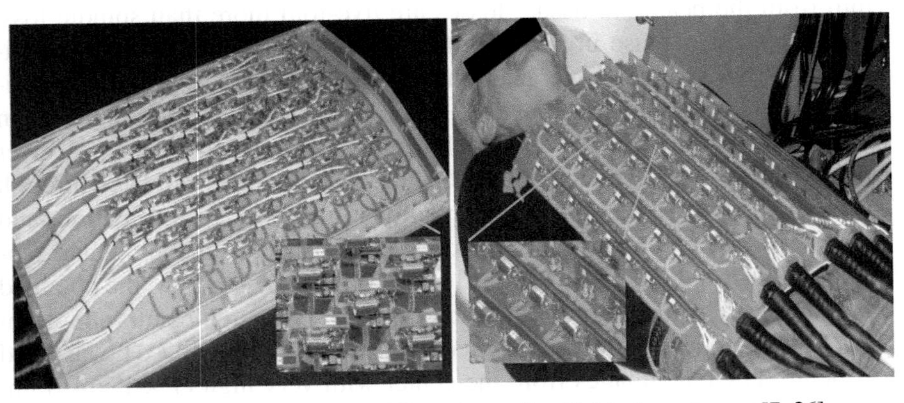

Figure 10.1 Flexible anterior RF receiver coil with 64-element array [7, 26].

developed by [6] using the initial idea of a monolithic resonator having a dedicated region of superficial imaging in human skin or in small animal imaging. Flexible thin polymer films were used as a dielectric substrate, to shape the micro-coil into non-planar surfaces, and to compare the results with a flexible 15mm diameter RF coil predicted to conduct a saline fantasy MRI proton. A maximum SNR gain of 2 is achieved using the coil developed over the phantom surface and the same was achieved by the plane RF coil. A high-density multi-bay MRI reception array was described by [7] which produces highly-accelerated parallel images. The array method consists of 2 clamshells that contain 64 anterior and posterior bends, resulting in an anterior array considering the reduction factor and 2D-3D pulse sequences.

The 19F/1H dual-frequency RF coil is built according to the universal strategy which allows for the development of many coil geometries [8] and its given in two stages. First, the coupling resonator is suited with good impedance for two harmonic oscillating modes. Second, 19F/1H dual nuclei imaging is observed with an electric test bench, the equivalent and homogeneous field distribution at 19F/1H frequencies were perceived in the test bench with fantastic pictures. The two standard prototypes of 19F/1H volume bobbins (Birdcage and Saddle) of 4.7T are designed, performed, and assessed. The sensitive and homogeneous architecture of the dual-frequency 19F/1 H coil in vivo mouse imagery was finally confirmed.

In [10] established functional MRIs (efMRI), based on spatial and temporary characteristics of the macro vasculature and neuronal-specific micro-vasculature (MRIs) signal (BOLDs), can be sensitive to the lamal- and columnar organization, however, studies in cortical architecture on this scale are uncommon. The spindle includes a very small $1 \times 2cm^2$ unit arranged in four flexible 4-element modules i.e. 16 channel elements, placed within 1mm from the head and tissue; losses are five-fold more than spindle loss, resulting in an increase in pre-amplifier disconnection. The BOLD's sensitivity for high spatial-temporal is 1mm isotropic with0.4s, whereas multiple-slice and echo planar acquisition is approximately 2.2 times higher than the typical 16-channel coil. [9] Established the 2D 16-channel transceiver MRI ranging at 7.0T. The RF protection approach has been validated through Specific Absorption Rate (SAR) simulations in the design, evaluation, and implementation process. 2D CINE FLASH images, T2^* mapping, and separation of fat-water images were used to perform Heart imagery, the obtained RF features are suitable for all the subjects. The simulation results show that SAR values obtained are well reduced and base SNR with 7.0T was used to obtain 2D CINE images from the center with very high $(1 \times 1 \times 4)$ mm^3 spatial resolution. Without greatly affecting the image quality, the proposed spool array supports 1D acceleration R = 4 factors. [23] developed a new method of integration, in which half of the spiral assembly can be positioned within the transducer acoustic processing volume and submerged in an acoustical coupling medium (generally degusted water), of the human brain processing system MRgFUS in the year 2014. The other half of the spindles volume assembly is connected with RF drive cables directly outside the transducer and results in uniform B1 in the heart therapy area, resulting in contact between the body spindle and the transducer. [11] Have built and evaluated an antenna monopoly array that can increase the sensitivity in the brain center by 7. A monopoly antenna array of 8 channels and a traditional 8-channel surface coil array have been assessed and compared for transmission properties of basic SAR and sensitivity. The sensitivity has been mapped by separating the SNR value through the flip angle distribution, resulting in a uniform sensitiveness of monopoly antenna array for the entire brain and 1.5 times increase in sensitivity increase in the center of the brain compared to the surface spiral array. [24] Describes the positioning of 31 MRI brain detecting coils on the PET detector ring in 2015; it absorbs and spreads photons to simultaneously acquire MR & PET pictures. The technique is used to remotely position the preamplifiers, coaxial size, bobbin size, and panel size to recognize PET / MR output barriers. The technique used to improve sensitivity in 1H coils is the use of close-in multi-element arrays using the scanner body RF spiral. As a result typical clinical field, strength has a low operating frequency of 23Na as becomes evident. The high-power RF Coil for 23Na MRI were developed [25] in brain and musculoskeletal applications.

10.2.3 Research development of industry version of MRI coils

Renowned flexible MRI coils with screen-prints will reduce the time taken for scanning. Light and flexible MRI coils have been developed, which produce MRI images of great quality and in future, it may result in shorter MRI scanning times. MRI scans will provide patient's life-saving data whereas other scans such as CT or PET doesn't provide. The patients who don't get affected by claustrophobia may also be uncomfortable while scanning even though coils are thin and flexible. Moreover, during the MRI scan large amount of noise is generated. The major issue, however, is that RMI devices can take more than an hour, or sometimes, to generate the photos the doctor wants and patients need to remain completely idle during the period. Such problems have to be overcome, making it difficult for patients, particularly pediatric patients. Doctors use anesthesia to help immobilize children, which is an additional danger for this vulnerable group of patients.

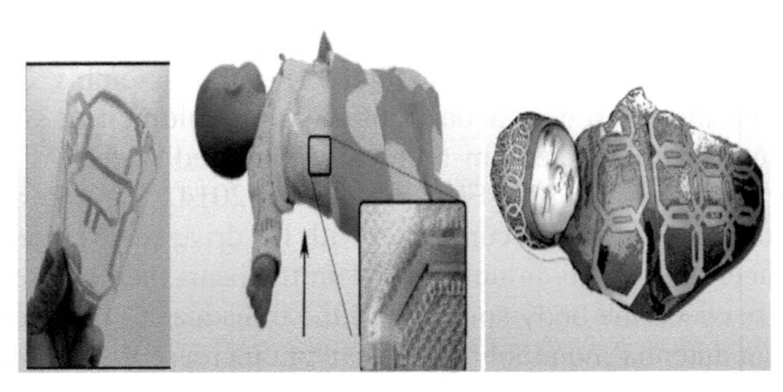

Figure 10.2 Left side of the figure shows an MRI scan coil separated from the patient and the right side of the figure shows that the coils are positioned against the patient with a significantly improved quality.

A new flexible MRI RF coils are developed to solve all the above problems [26]. Screen printers can be individually made for patients of different sizes, like babies or children, or individually adapted to patients as required. Screen-printing is the process that is used for printing T-shirt designs. Thanks to their lightness and versatility, these coils may be adequately wrapped on all sides of the patient's body, which improves the sensitivity and clarification of images.

The spools can be reused and printed directly for each patient rather than using the MRI machines currently using 1.5T and 3T. In a prototype, a blanket has been developed with the spools inside, where a child will be wrapped. This new concept is to be made feasible for mass production in a near future and could provide a cost-effective way for doctors to access the information they need, with the least amount of patient distress, with increasingly popular printing technologies. Greatly enhancing patient satisfaction at a reasonably lower cost. Regular MRI scanning causes damage whereas the flexible MRI coils are advantageous for pediatric patients and parents. The latest evolution is a formidable example of the upcoming technology which is integrated with a better experience for the patients.

10.3 Proposed Methodology

10.3.1 A design a coils by meta-materials

A number of novel approaches for clothing, dispersal, and absorption have been introduced in recent years based on metasurfaces and meta-materials. Many flexible technical protocols based on metasurfaces and metamaterials are available still, future work has several problems to resolve. First is to reduce the fundamental resonator dimension to a dimension as small as or much smaller than the size of a deep sub-wavelength [27]. Second is to overcome the larger thickness and huge size of profiles by understanding the working principle with low microwave frequencies. The meta-materials have a stacked EBG structure consisting of two arrays of metal patches with each other's diagonal offset. With vertical metal plated vias, the top layer is linked to the metal back of the dielectric substrate while the lower layer floats. Designing and characterizing of EBG arrangement was simulated in HFSS full-wave simulator using the FEM algorithm consisting of periodic boundary conditions (PBC) and EMPIre XCcel FDTD simulator was used to analyze the finite structure. EBG arrangement along with meander dipole coil is considered for analyzing the geometry of the proposed finite structure [28]. The RF coils doesn't have uniformity in the field because the length is independent and greater than $0.25\lambda_g$. In order to attain field uniformity, Zeroth Order Resonator (ZOR) metamaterial having steady composite right hand left (CRLH) is selected instead of RF coil element [29]. In the transmission line TEM coils, the metamaterial unit cells are initiated to form a head coil for a 10.5 T system. Full-wave simulation at 447.06 MHz, the field B1 is uniform when meta-cells are introduced into the spindles. In a broader area compared to conventional, the proposed meta-TEM head coil displays a greater magnetic field with high efficiency [30]. The EBG metamaterial is used for designing and the results were obtained are numerically optimized for 7T MRI applications. The results obtained using an EBG arrangement and analyzed the achievable insulation level for numerical simulations and bench measures of the array spool setting. The experimental results of RF magnetic Fields and Sar Patterns were numerically justified the effect of decoupling [31]. The metamaterial-based square loop coil element is introduced first, followed by a simulation evaluation of the proposed 10.5T system. In the end, the extension of the metamaterial line not possible with any of the other current designs is shown [32]. The effect of transmission efficiency in bird cages is assessed by comparing the numerical value of flat and volumetric metasurface based SAR of the wireless coils. Human body's realistic model is used for the comparison [33]. The problem is to determine how an inter-digital capacity metasurface unit cell can improve the performance of the SNR [34]. Compact, easy to manufacture Hilbert-based metasurface resonator, which increases effectively the sensitivity of the 7T MRI scanner using RF field intensity and the absorption depth of the RF belt [35]. The first application in volume coil design is demonstrated by the artificial dielectric application. The artificial dielectric resonator works as an electromagnetically combined passive wireless structure with a commercial

3T clinical MRI body coil and is compared with the conventionally bigger dielectric resonator [36]. It demonstrates experimentally that a 1.5T MRI scanner increases the local transmission efficiency using a metasurface made up of a series of brass wires integrated into a medium of low loss for high permittivity. The placement of an RF field produced alongside the body coil in such a structure within the scanner is strongly linked to the lowest frequency electromagnetic proprietary mode of metasurface [37]. The broad range of hetero nuclear experiments are done using a double coil system set-up, which consists of two independent coils. One of the coils is tuned for a frequency of ^{1}H for getting the anatomical images for reference. A metamaterial-inspired coil is used for a wide range of frequencies by X-nuclei, where frequencies ^{2}H and ^{31}P are selected as the frequency of margin. The two nuclei Larmor frequency for 11.7T has a frequency gap of 127.7 MHz [38]. A hybridized meta atom having four wires is designed and constructed. It is surrounded by a lower half-cylinder having a diameter of 2.5cm for rat brain imaging. The construction of the above setup is compared with numerical design and experimental values with the bird cage coil using a preclinical 17.2T MRI scanner. It is explored that Vivo rat brain imaging structure has more benefits [39].

10.3.2 Implementation using big data digitization analysis through wireless networks

The receiver was shown to be linear above the range of 84 dB in the dynamic measurement range. This includes both analog and sampling effects as well as digital filtration effects. With 14-bit ADC, a sampling factor of 40 for a noise size is about 2 bits, the digitalization effects appear to limit the dynamic range almost entirely. The volume of new data produced through40 min of f-MRI acquisition of data to be more than 40 GB per object with a data performance of approximately 20 MB/s for an ideal 16-channel f-MRI evaluation (1 s TR, 8 slices, 128 × 96 matrix size, and 2 μs space sample). The new data flow is very large and data management and storage are a challenge [40]. Integrating the RF receiver with the present CMOS technology is essential to reduce the recipient size and place the receiver on receiving spindles, thereby digitalizing the data obtained. This makes it possible to use optical fibers to transmit data, avoiding coaxial cables with substantial cost benefits and usability. Wearable, adaptive detector arrays for measurements with 3T and 7T MRI is fully equipped with integrated CMOS receivers. To reduce power and lighten the need for A/D conversion, Application-specific integrated circuit (ASIC) MRI receiver directly converts instead of getting direct samples

[41]. The system uses RF to reduce the effect on the MR picture through the time-consuming Wireless Power Transfer (WPT) system. A 10 mF condenser is used to supply the power to the on and off system continuously. As wireless spindles would take up to 100-300 mW per channel, 10 Watt would be required for a 32-channel array. This shows a Wi-Fi transmission of up to 11W for this type of system [42]. The on-coil digitalization for RF signals has the dynamic range DR ≥ 16-bit and different compressed data rate method requires RF signals from modern MRI settings (e.g. 3T, 64 RF Receiver Channels) for data rates > 500Mbit/s. 60 GHz Wi-Gig and optical wireless communication seem to suit the scheme for wireless-mobile MR data transmission; however, it is still necessary to verify the on-coil features of the MRI scans. In addition to RF signals, control signals for on-coil components must be managed, including active detuning, MR system synchronization, and B_0 shimming. Wireless power supply is becoming significant with a wide range of other on-coil components in particular. In contrast to huge MR compatible batteries and energy recovery with low power output, WPT systems greater than 10W appear an appealing option. The system has fully wireless RF coils and has eventually enhanced realism for the future by incorporating effective schemes for modern scientific advancements. Innovation in wireless technology, MR compatibility, and wireless power supply is especially required besides continuous improvement of all three subsystems [43].

10.3.3 To design a flexible adaptive multituned RF coils

A new RF coil for use with MR imaging in radiation therapy patients, is flexible and highly decoupled. Coil performance was assessed using two coils (Sns) and four-element (noise-relationship) element combinations, based on the overlap distance of the coil, and the comparison between designed values and the values achieved by conventional coil elements was analyzed. Coils were calculated by means of SNR and noise-relationship measuring. The RF coils will be integrated with highly flexible substrates that fit the patient precisely. This design discusses the disadvantages of traditional surface coils and potentially impacts MR imaging for diagnostic as well as for radiation therapy [44]. Develop a mechanically adjustable 3T MRI RF array to overcome the design model by an adjustable but rigid coil design and the case for easy use in voluntary or treatment studies. The designed array is suitable for various vital organ applications such as brain and muscular skeleton, where sensitivity to detection and comfortable for a patient to easily adapt the increase in array to each anatomy. Moreover, the parallel imaging performance should be improved by enough channels in two directions [45]. MRI guided radiotherapy array proposes a treatment for patients which considers the design and feasibility of thick high impedance coils (HICs) to enhance the imaging performance of 1.5 T MR-linac for on-body placements. The anterior element is flexible to closely match all patients' body contours. The rear element for the optimal imaging sensitivity is placed right beneath the patient. Up to 32 channels can be used with the full range. The configuration and size of the supporting materials and conductors are optimized for minimizing the radiation therapy impacts in all respects. Ideally, it has a minimum impact. The designed method of the array does not include the treatment planning system (TPS) or position tracking with respect to time dynamically and reduces the use. In the end, functional prototypes of one channel and five channels are manufactured and the gain of SNR is quantified from our on-board approach. The current clinical array has been activated both with and without a beam of radiation [46]. The improvement of SNR includes a formulation of the bow array that includes a universal system for the part of the body, but it's regularly rigid and involves a certain positioning of the patient. Whatever it's positioning, the flexible and extendable Purdue coil can be placed in an area or joint near the skin [47]. The multi-tuned coils design concept, is mainly for brain applications. The analysis is divided into two parts to guide readers: state-of-the-art based on single or multiple design schemes and modern technologies. Additional information about the detailed design methods especially double-tuned coils using traps, PIN diodes, nested and meta-materials are described in detail with descriptions of their novelties, optimum results, and trade-offs are provided in each subsection [48].

10.4 Conclusion

The various fabrication techniques of MRI coils with different applications are presented in this survey including the transceiver operational performance, development of recent technology, etc. Currently, a multi-dimensional array of MRI coils developed in the wearable features; methodology and their advantages are implemented for a specific application. Finally, the prototypes of MRI coils are designed for the following research dimensionality, such as less Specific Absorption Ratio (SAR) for frequency of 10 GHz, flexible, foldable, SNR reduction, and wearable in body-centric application. These design parameters are considered for the development of MRI coils using composite materials.

References

[1] Lauterbur PC et al, 'Image formation by induced local interactions: Examples employing nuclear magnetic resonance Nature', 242, pp.190–191, 1973.
[2] Qiang Zhang et al, 'A Multielement RF Coil for MRI Guidance of Interventional Devices', Wiley-Liss, Inc. Journal of Magnetic Resonant Imaging, 14(5), pp.56–62, 2001.
[3] Xiaoliang Zhang, Kamil Ugurbil and Wei Chen, 'Micro-strip RF Surface Coil Design for Extremely High-Field MRI and Spectroscopy', Wiley Liss. Inc. Magn Reson Med, 46(6), pp.443–450, 2001.
[4] Xiaoliang Zhang, Xiao-Hong Zhu, and Wei Chen, 'Higher-Order Harmonic Transmission-Line RF Coil Design for MR Applications', Wiley-Liss. Inc. Magn Reson Med, 53(4), pp.1234–1239, 2005.
[5] Robert G. McKinnon and Ravi S Menon et al, 'SENSE Optimization of a Transceiver Surface Coil Array for MRI at 4 T', Wiley-Liss. Inc. Magn Reson Med, 56(5), pp.630–636, 2006.
[6] M. Woytasik J.C. Ginefri J.S and Raynaud M et al., 'Characterization of flexible RF micro coils dedicated to local MRI', Springer-Verlag, Microsystems Techno, 13(4), pp.1575–1580, 2007.
[7] Christopher J. Hardy. et.al. '128-Channel Body MRI With a Flexible High-Density Receiver-Coil Array', Wiley-Liss, Inc. J. Magn. Reson. Imaging, 28(5), pp.1219–1225, 2008.
[8] Lingzhi Hu, MS. Frank D. Hockett, MS. and Junjie Chen. DSc et al, 'A Generalized Strategy for Designing 19F/1H Dual-Frequency MRI Coil for Small Animal Imaging at 4.7 Tesla', Wiley-Liss, Inc. J. Magn. Reson. Imaging, 34(6), pp.245–252, 2011.
[9] Christof Thalhammer. Wolfgang Renz et al, 'Two-Dimensional Sixteen Channel Transmit/Receive Coil Array for Cardiac MRI at 7.0 T: Design, Evaluation, and Application', Wiley Periodicals Inc. J. Magn. Reson. Imaging, 36(9), pp.847–857, 2012.
[10] N. Petridou. et al, 'Pushing the limits of high-resolution functional MRI using a simple high-density multi-element coil design', John Wiley & Sons Ltd. NMR Biomed, 2012.
[11] Suk-Min Hong. Joshua Haekyun Park. And Myung-Kyun Woo et al, 'New Design Concept of Monopole Antenna Array for UHF 7T MRI', Wiley Periodicals, Inc. Magn Reson Med, 71(7), pp.1944–1952, 2014.
[12] McRobbie DW. Moore EA. and Graves MJ et al, 'MRI from picture to Proton', 2nd Ed. New York: Cambridge University Press, 2006.

[13] Wright SM, 'Receiver loop arrays', Encyclopaedia of Magnetic Resonance, pp.1–13, 2011.

[14] Barrie Smith N, Webb A, 'Introduction to medical imaging Physics, Engineering and clinical applications', Cambridge UK: Cambridge University Press, 2010.

[15] Mispelter J, Lupu M, and Briguet A, 'NMR probe heads for biophysical and mio-medical experiments: theoretical principles & practical guidelines', London: Imperial College Press, 2006.

[16] Chen CN, Hoult DI, and Sank VJ, 'Quadrature detection coils A further 2 improvement in sensitivity', J Magn Reson, 54(2), pp.324–327, 1983.

[17] Glover GH, Hayes CE, and Pelc NJ et al, 'Comparison of linear and circular polarization for magnetic resonance imaging', J Magn Reson, 54(14), pp.255–270, 1985.

[18] Ben G. Lawrence et al, 'A Time-Harmonic Inverse Methodology for the Design of RF Coils in MRI', IEEE Transactions On Biomedical Engineering, 49, pp1, 2002.

[19] Jing Fang. M. S. Chow. and K. C. Chan et al, 'Design of Superconducting MRI Surface Coil by Using Method of Moment', IEEE Transactions On Applied Superconductivity, 12, pp.2, 2002.

[20] Stephen G. Worthley et al, 'A Novel Non-obstructive Intravascular MRI Coil In Vivo Imaging of Experimental Atherosclerosis', American Heart Association Inc, 23(3), pp.346-350, 2003.

[21] Christopher J. Hardy et al, 'Large Field-of-View Real-Time MRI with a 32-Channel System', Wiley-Liss, Inc. Magn Reson Med, 52(5), pp.878–884, 2004.

[22] Christopher J. Hardy et al, '32-Element Receiver-Coil Array for Cardiac Imaging, Wiley-Liss, Inc. Magnetic Resonance in Medicine, 55(6), pp.1142–1149, 2006.

[23] Ronald D Watkins et al, 'Integration of an Inductive Driven Axially Split Quadrature Volume Coil with MRgFUS System for treatment of Human Brain', Proc. Intl. Soc. Mag. Reson. Med, 22, 2014.

[24] Christin Y. Sander et al, 'A 31-Channel MR Brain Array Coil Compatible with Positron Emission Tomography', Magn Reson Med, 73(6), pp.2363–2375, 2015.

[25] Graham C. Wiggins et al, 'High Performance RF Coils for 23Na MRI: Brain and Musculoskeletal Applications, 'NMR Biomed', 29(2), pp.96–106, 2017.

[26] Corea, J. R. et al, 'Screen-printed flexible magnetic resonance imaging receives coils', Nat. Commun, 7, pp.10839, doi:10.1038/ncomms10839, 2016.

[27] Yafan. and Jiafu Wang et al, 'Recent developments of metamaterials / meta-surfaces for RCS reduction', EPJ Appl. Metamaterials, 6, pp.15, 2019.

[28] Gameel Saleh et al, 'EBG Structure to Improve the B1 Efficiency of Strip line Coil for 7 Tesla MRI', 6th European Conference on Antennas and Propagation (EUCAP) IEEE, doi.org/978-1-4577-0919-7/12, 2011.

[29] Vijayaraghavan Panda et al, 'Zeroth Order Resonant Element for MRI Transmission Line RF Coil', IEEE, doi.org/978-1-5090-2886-3/16, 2016.

[30] Shao Ying Huang, 'Metamatarial-inspired Transverse Electromagnetic (TEM) Head Coil at 10.5 T', IEEE, doi.org/978-1-4673-9811-4/16, 2016.

[31] Anna A. Hurshkainen et al, 'Element decoupling of 7 T dipole body arrays by EBG metasurface structures: Experimental verification', Journal of Magnetic Resonance,; 269: 87–96. 2016.

[32] Vijayaraghavan Panda et al, 'Metamaterial Loop Body Coil Element for 10.5T MRI', IEEE, doi.org/978-1-5386-3284-0/17, 2017.

[33] Alena V. Shchelokova et al, 'Metasurface-Based Wireless Coils for Magnetic Resonance Imaging', IEEE International Conference on Microwaves, Antennas, Communications and Electronic Systems (COMCAS), doi.org/978-1-5386-3169-0/17, 2017.

[34] Tingzhao Yang et al, 'A Single Unit Cell Metasurface for Magnetic Resonance Imaging Applications', 12th International Congress on Artificial Materials for Novel Wave Phenomena –Metamaterials, doi. org/978-1-5386-4702-8/18, 2018.

[35] Elizaveta Motovilova. and Shao Ying Huang, 'Hilbert Curve-Based Meta-surface to Enhance Sensitivity of Radio Frequency Coils for 7-T MRI', IEEE Transactions On Microwave Theory and Techniques, doi. org/0018-9480, 2018.

[36] Anna A. Mikhailovskaya et al, 'A new quadrature annular resonator for 3 T MRI based on artificial-dielectrics', Journal of Magnetic Resonance, 291, pp.47–52, 2018.

[37] Alena V. Shchelokova et al, 'Experimental investigation of a meta-surface resonator for in vivo imaging at 1.5 T', Journal of Magnetic Resonance, 286, pp.78–81, 2018.

[38] V.A. Ivanov et al, 'RF-coil with variable resonant frequency for multi-hetero nuclear ultra-high field MRI', Photonics and Nanostructures – Fundamentals and Applications, doi.org/S1569-4410(19)30139-7, 2019.

[39] Marc Dubois et al, 'Enhancing surface coil sensitive volume with hybridized electric dipoles at 17.2 T', Journal of Magnetic Resonance, 307, pp.106567, 2019.

[40] Jerzy Bodurka. and Patrick J. Ledden et al, 'Scalable Multichannel MRI Data Acquisition System', Magnetic Resonance in Medicine, 51(5), pp.165–171, 2004.

[41] Benjamin Sporrer et al, 'Integrated CMOS Receiver for Wearable Coil Arrays in MRI Applications', EDAA, doi.org/978-3-9815370-4-8, 2015.

[42] Kelly Byron et al, 'An RF-gated wireless power transfer systems for wireless MRI receive arrays', Concepts Magn Reson Part B Magn Reson Eng., 47B(4), doi:10.1002/cmr.b.21360, 2017.

[43] Lena Nohava. and Jean-Christophe Ginefri et al, Perspectives in Wireless Radio Frequency Coil Development for Magnetic Resonance Imaging', Front. Phys, 2020.

[44] Kiaran P McGee. and Robert S Stormont et al, 'Characterization and evaluation of a flexible MRI receive coil array for radiation therapy MR treatment planning using highly decoupled RF circuits', Phys. Med. Biol', pp.63, 2018.

[45] Roberta Frass-Kriegl et al, 'Flexible 23-channel coil array for high resolution magnetic resonance imaging at 3Tesla', PLOS ONE, https:// doi.org/10.1371/journal.pone.0206963, 2018.

[46] Stefan E Zijlema et al, 'Design and feasibility of a flexible, on-body, high impedance coil receive array for a 1.5 T MR-linac', Phys. Med. Biol, 64, 185004 (13pp), 2019.

[47] Jana Vincent and Joseph V. Rispoli, 'Stretchable, wearable coils may make MRI, other medical tests easier on patients', IEEE Transactions on Biomedical Engineering, 2020.

[48] Chang-Hoon Choia et al, 'The state-of the-art and emerging design approaches of double-tuned RF coils for X-nuclei, brain MR imaging and spectroscopy', A review Magnetic Resonance Imaging, 72(12), pp.103–116, 2020.

Chapter 11

Wireless Sensing Based Solar Tracking System Using Machine Learning

S. Saroja[1], R. Madavan[2], Jeevika Alagan[3], V.S. Chandrika[4], and Alagar Karthick[4*]

[1]Assistant Professor/ Information Technology, Mepco Schlenk Engineering College, Sivakasi, Tamilnadu, India
[2]Associate Professor & Head Department of Electrical and Electronics Engineering, PSR Engineering College, Sivakasi, Tamilnadu, India
[3]Assistant professor, PG and Research department of chemistry, Thiagarajar college, Madurai-625009, Tamilnadu, India
[4]Associate Professor, Renewable Energy Lab, Department of Electrical and Electronics Engineering, Coimbatore-641404, Tamilnadu, India
Email: saroja@mepcoeng.ac.in; srmadavan@gmail.com; jeevichemist@gmail.com; chandrika.vs@kpriet.ac.in,
Corresponding author Karthick.power@gmail.com

Abstract

Solar energy is a clean source of energy that falls under the category of renewable and sustainable energy and is widely accessible around the globe. In solar power systems, solar tracking systems are critical for optimizing energy output from the sun. Single and dual-axis solar trackers have traditionally been used to move solar panels in different directions based on the sun's beams in order to enhance energy. To optimize the energy gain, the deployed tracker must actively follow the sun's rays and adjust its location appropriately. The key components needed for constructing tracking systems are sensors, microcontroller-controlled control circuits, and servomotors with supports and mountings. Two servo motors are used to adjust the position of the solar panel so that the sun's beam remains aligned with the solar panel. The suggested solar tracking system, which is based on machine learning, allows the solar panel to spin in any direction. The proposed machine learning-based classification technique examines and categorizes sensor outputs as defective or not. If the sensor results are awarded the class label "Non-faulty," the servo motor adjusts the position and direction of the solar panel depending on the input sensor data. If the sensor readings are classified as "Faulty," the servo motor will modify the position and direction of the solar panel using a regression-based machine learning technique.

11.1 Introduction

Energy has become a critical component in all industries as a result of the massive Industrial Revolution. Energy use is also rising on a daily basis, resulting in depletion of energy resources and scarcity of resources. Renewable energy, often known as clean energy, is the most efficient energy source since it is derived from naturally renewing resources such as the sun and wind. Solar and wind energy generating are the two most rapidly growing industries in the clean energy world today. The major advantages of employing renewable energy sources are lower carbon and other pollution emissions. Solar and wind energy are at the top of the priority list for different renewable energy sources. Solar energy has been used by humans for crop cultivation, clothing drying, and food production for thousands of years. Solar energy is being utilized to generate electricity. Photovoltaic (PV) cells are used in solar energy systems to convert sunlight into electricity. The power generated by direct lighting of PV cells' surfaces is maximized. This is accomplished by monitoring the sun's rays and altering the surface of the solar panel, which is comprised of PV cells, in response to the sun's rays. This chapter discusses how to detect sun rays using Light Dependent Resistor (LDR) sensors and how to use machine learning techniques to learn about the environment and adjust the solar panel's surface appropriately.

11.1.1 Purpose

The suggested solar Tracking system in this chapter use LDR to detect sun rays and hence the sun's direction. The solar panel moves in time with the rising sun by utilizing the LDR output. As a consequence, the solar panel saves energy from the sun in both directions. When compared to existing fixed panels, the suggested design allows for more energy to be extracted from the panel. The suggested Solar Tracking system is easy to set up and requires minimal modification to the panel installation. It has the appearance of a standard solar panel, with the addition of LDR for monitoring the sun's direction.

11.1.2 Overview

The proposed Solar Tracking system deals with the tracking of solar energy in both direction from the sun and saves it and converts it into electrical energy continuously. The main challenge lies in tracking solar energy from both the direction of the sun. The procedure is to simply fix the LDR in both sides of the solar panel; the LDR, on turn, rotates the solar panel based on the direction of the sun and gains energy. Finally, we can get solar energy from both directions and hence can produce more renewable energy.

11.2 Solar Tracking System

This section deals with the architecture and main components of the solar tracking system.

11.2.1 Architectural description

The configuration requires mounting two LDRs on both sides of the solar panel, as well as a servo motor on the bottom that spins the solar panel. The servo motor will aid the solar panel in moving towards the LDR with the lowest resistance, i.e. the LDR on which light is falling; in this manner, the solar panel will always follow the light. If both LDRs receive the same quantity of light, the servo motor will not move the panel, and the solar panel's position will remain unchanged. The direction of the rotation is decided by two algorithms: The Clustering algorithm and the Edge Detection algorithm. Dual-axis solar tracker is created using 2 LDR's and a servo motor to manage the

Figure 11.1 Solar panel assembly.

Figure 11.2 Main components of solar tracking system.

sun angle change over the years. The mean resistance across all the horizontal axis and vertical axis is calculated and it is fed as input for the algorithms finally, the rotation angle is obtained.

11.2.2 Main components

The major components of the Solar Tracking System are as follows:

- Servo Motor
- LDR
- Solar Panel

11.2.2.1 Servomotor

A servo motor is an electronic device that rotates a connected item more precisely in terms of angle, acceleration, and velocity. To regulate the rotation of the basic motor, a servo mechanism is used. If the motor is powered by DC, it is referred to as a DC servo motor, and if it is driven by AC, it is referred to as an AC servo motor. The servo motor is utilized in many industrial and electrical applications where automatic rotation is important, such as toy cars, robots, and kiosk machines, due to its simple operating concept.

The servo motor's operation is based on the Pulse Width Modulation theory (PWM). The angle of rotation is controlled by the duration of the pulse. The rotation and position of a servo motor are controlled by a closed-loop system that employs feedback. The needed position is provided as an input to the servo motor, which includes encoder devices that offer feedback on position and speed. After that, the output's measured real location is compared to the projected position. An error signal is issued if the actual output position varies from the intended location. It directs the motor to spin in the proper direction in order to bring the output shaft to the proper position. The outcome is represented in the error signal, which stops the motor when the actual output position approaches the predicted position.

The servo motor is employed in the proposed work to spin the solar panel towards the LDR based on the resistance value. The solar panel's position will be altered by rotating it towards the LDR with the least resistance, and therefore towards the direction of the sun's beams. If both LDRs detect the same quantity of sunshine, the location of the solar panel stays unchanged. Based on the resistance value provided by LDR, the servo motor spins the solar panel. The primary goal of using the servo motor is to direct the solar panel into lower resistance LDR.

11.2.2.2 LDR

An LDR, or light-dependent resistor, is a kind of resistor that has a variable resistance value depending on how much light is reflected on its surface. LDR is also known as a photoresistor, photocell, or photoconductor. In the absence of light, a typical light dependent resistor has a resistance of 1 MOhm, while in the presence of plentiful light, it has a resistance of a couple of KOhm. The resistance value drops as light shines on the resistor. The basic feature of LDR is the change in resistance value as a function of the quantity of light. This quality makes LDR a good choice for detecting the presence of light in electric circuits such as automated street light controllers, fire alarms, smoke alarms, light-activated switches, automatic night security lights, automatic emergency lights, and so on. For example, LDR is used in smart street lighting to switch on and off lights on their own. Cadmium sulphide (CdS), a photoconductor with no or few electrons when not illuminated, is the main component driving the LDR. In the dark, it possesses a high resistance in the Mega ohm region. When light shines on the sensor, it illuminates the electrons, increasing the conductivity of the material. The photons absorbed by the semiconductor supply enough energy to the band electrons to enable them to jump into the conduction band when the light intensity surpasses a certain frequency (threshold value). As a consequence, the liberated electrons or holes conduct current, reducing the resistance significantly (to 1 Kilo ohm). LDR is divided into two types. The following are the details:

Intrinsic Photo Resistors are photo resistors that are built into the device itself.

In the production of intrinsic photo resistors, pure semiconductor components such as silicon or germanium are employed. When light strikes the surface of these semiconductor materials, the electrons in the valence band are stimulated and transported to the conduction band, resulting in an increase in the number of charge carriers.

Photo Resistors (Extrinsic)
Impurities are doped into extrinsic photo resistors. Because of the existence of impurities, new energy bands from above the valence band. This freshly formed energy band is densely packed with electrons. As a result, the space between the bands narrows, requiring less energy to transfer electrons between them.

LDR is a feasible choice in a solar tracking system to detect the existence of sun rays due to its inexpensive cost and easy functioning concept.

11.2.2.3 Solar panel

Solar panels, also known as PV panels, are devices that convert light energy (derived from the sun's rays) into electrical energy, which may be used to power equipment in the house, business, and so on. Solar panels are made up mostly of solar cells. Through the photovoltaic effect, solar cells, also known as PV cells, transform light energy into electric energy.

The photovoltaic effect is a mechanism in which a PV cell generates an electric current or voltage when exposed to sunshine. In the year 1839, Edmond Becquerel developed the photovoltaic effect. The photovoltaic effect is used by solar cells to transform light energy into electric energy. A p-type and an n-type semiconductor are used in these solar cells. A p-n junction is formed when these semiconductors are linked together. An electric field is established in the region of a junction formed by joining these two kinds of semiconductors when electrons flow to the positive p-side and holes flow to the negative n-side. As a consequence of this field, negatively charged particles flow in one direction while positively charged particles flow in the other way.

Photons are light's atomic elements, which are electromagnetic energy bundles. When light rays strike a photovoltaic cell, the photons are absorbed by the cell. When these cells are exposed to light rays of an appropriate wavelength, the energy from the light ray is transmitted to one atom of the semiconducting material present in the p-n junction. The energy from the light wave is transferred to the electrons in the semiconductor material. As a result, electrons move from the valence band to the higher-energy conduction band, creating a "hole" in the valence band. This journey of the electron results in the formation of an electron-hole pair, which consists of two charge carriers. Electrons cannot travel in the unexcited state because they create bonds with the surrounding atoms. In the excited state, electrons in the conduction band are free to flow through the material. The electric field created by the p-n junction causes electrons and holes to flow in opposing directions. The available free electrons are pulled to the n-side. As a consequence of this movement, the cell creates an electric current. After the electron has shifted, there is a "hole" that remains. This hole has the potential to shift to the p-side. The Photovoltaic Effect refers to the whole process.

Characteristics (2.3)
The planned solar tracking system's key selling point is its inexpensive price. The Solar Tracking System is inexpensive since it only requires a few low-cost components in addition to a standard solar panel. To obtain energy from opposite directions from the sun, two LDR are built on each side of the solar panel. The servo motor is the component that rotates the solar panel in response to the LDR's reaction.

11.2.4 Limitations

The only stumbling block in the implementation of the solar tracking system is determining the solar panel's orientation. If the solar panel is pointed in the right direction, it will produce the most amount of electricity. For optimal efficiency, the solar panel should be installed in either an east or west orientation. The solar panel's efficiency is increased even further by monitoring the sun's beams in both directions.

11.2.5 Dependencies and assumptions

Because the proposed solar tracking system is embedded in nature, its proper operation is contingent on the software and hardware being in working order. The software that has been created should be compatible with the hardware that has been selected. The hardware and software requirements should be worked properly, along with any necessary updates.

11.2.6 Specifications for requirements

This section provides a full discussion of the solar tracking system's numerous needs as well as all of its available capabilities.

11.2.6.1 Requirement for external interface

This section covers all of the solar tracking system's inputs and outputs, as well as the many software and communication interfaces and primitive user interface prototypes. The technician installing the solar panel must verify the LDR's state, and the two LDR's must be installed in different directions. Two LDRs must be in one direction and the other two must be in the other direction in a dual-axis tracking system. On the eastern side, for example, one LDR must be installed in the east and the other in the north-east. To preserve the whole angle of sun rotation on the western side, one LDR must be positioned in the west and the other in the south-west.

Hardware Interfaces: The solar tracking system will be installed, with all necessary components linked, and stored on the environment's terrace. The solar panel with a single-axis tracking system is rotated in just one direction (horizontal rotation is alone possible by servo motor output, which is fixed in alignment with the axis on the terrace). A servo motor aligned with the horizontal setup is mounted with the vertical axis in a dual-axis tracking system (ensure proper connectivity is maintained, as poor connectivity among the hardware components leads to malfunctioning of the component and also ends in hardware fault). Initially, all LDRs are kept at zero resistance.

Software Interfaces: The solar tracking system will be programmed using Arduino, and the solar panel's functionality will be tested by connecting it to a voltmeter.

Communication Interfaces: The Solar Tracking System does not need any end-user coordination or communication. As a result, no human interaction is required since the sole means of communication is via light rays incidence between the sun and the solar panel.

11.2.6.2 Requirements, both functional and non-functional

This section contains a full overview of the solar tracking system's functional needs as well as all of its features.

- The device should be installed on an open terrace with no obstructions to sunlight.
- The effectiveness of the solar tracking system may be enhanced by using dual-axis tracking instead of single-axis tracking; it is pleasant if there is no congestion in putting the device.

Requirements for performance: The allowed reaction time and throughput for effective system operation are referred to as performance requirements. The initial loading time for hardware configuration is 2 minutes, and subsequent answers are expected to take less than two seconds. When the light resistance is detected in all of the LDRs at the same time, servo rotation may occur in both the horizontal and vertical directions.

Design Constraints: The Solar Tracking system is a stand-alone program that runs on Arduino. Other microcontrollers may potentially be used to create this system. The related code may then be connected to the equipment in the Android environment via wifi or any other acceptable communication channel.

Standard Compliance: The overall design is consistent, and the execution follows the coding standards. In the upgraded real-time version, the system's graphical user interface is meant to have a consistent appearance and feel.

System reliability is determined by the various hardware components employed in the system's construction.

After installation and configuration, this system will be accessible 24 hours a day, seven days a week, 365 days a year. However, the system's functioning may be witnessed throughout the day.

Security: To manage the operation safely, the proposed Solar tracking system and its subsequent implementations, as well as communication interfaces such as Wi-Fi, are password secured.

The Solar Tracking System application is portable since it runs on an Android platform. Although it is connected to the Wi-Fi module in real-time, it must be close by for optimal functionality.

11.3 Machine Learning Algorithms

Machine learning is a new science in the computer engineering field, which adds a new capability to computers by making computers learn by themselves instead of the traditional way of programming. Machine learning makes the computer act autonomously without the intervention of human beings.

Machine learning is applied in a wide variety of fields including:

- Fraud Detection
- Recommendation systems
- Healthcare Systems
- Traffic Analysis
- Stock market prediction
- Biometric Recognition

- Pattern Recognition
- Spam Detection
- Autonomous Vehicles
- Weather pattern prediction
- Translation systems
- Natural language processing

Machine learning algorithms are broadly classified into three different groups based on their learning style. They have supervised learning, unsupervised learning, and semi-supervised learning. The proposed solar tracking system uses two supervised learning algorithms namely classification and regression for increasing the energy gain.

11.3.1 Supervised learning

Supervised learning algorithms represent input-output linkages and dependencies in such a way that they can anticipate the output values for new data based on the relationships learned from prior data sets. In this technique, computers are fed with the training data containing input and output combinations by a human being. Computers learn the pattern and the possible relationship that exist between input and output and create a model out of it. The training process continues until the model has learned completely from the training data. The completeness of the learning is measured by the accuracy of the model. Supervised learning techniques can be further divided into two categories namely: Classification and Regression

11.3.1.1 Classification

Classification is defined as the process of classifying the given input data into different output classes. The main task of any classification algorithm is to assign or predict the class label (from the set of available labels) for the given input data. For example, in the case of a spam filtering application, given an email, the classification algorithm has to assign the class label as either "spam" or "no-spam". That is the given email is classified as either spam or ordinary mail.

Mathematically, classification is expressed as the process of approximating a mapping function (f) from input variables (X) to output variables (Y). Classification comes under a supervised machine learning technique because in the training phase the input data set is fed along with the class labels to train the system.

The above-said example classification problem for email spam detection involves only two class labels, "spam" and "no-spam"; hence this is of type binary classification.

To implement this classification, the classification model has to be trained with a sufficient number of samples. Hence some known, "spam" and "no-spam" emails would be used as the training data. After completion of the training phase, the newly built classification model can be tested with an unknown, email and the model reports the class label for the given unknown email as either "spam" or "no-spam". Similar to this binary classification, we can also have a multi-label classification, where the output labels are more than two.

The proposed Solar tracking system uses a binary classification algorithm to classify the LDR results as **"Faulty" or "Non-Faulty"**. As the system is having only two class labels, this falls under the group of a binary classification problem.

The following lists the different classification algorithms:

- Decision Tree Classifier
- K-Nearest Neighbour Classifier
- Random Forest
- Support vector Machine
- Naïve Bayes Classifier

Decision Tree Classifier: Decision tree classification falls under predictive analytics and uses an algorithmic way of splitting the given input data set into different classes based on certain conditions. It is one of the widely used most familiar classification approaches under the supervised learning techniques. The output variable is a class label, which will be assigned to the input vector based on certain conditions. Apart from classification, the decision tree algorithm is also useful for doing regression.

K-Nearest Neighbour Classifier: K-Nearest Neighbour algorithm can be used for solving the classification and regression under the Supervised Learning technique. K-NN algorithm uses similarity measurement as the main criterion for classification. For the newly arrived input sample, it measures the similarity between the new samples with all the available samples, and then a class label is assigned to the new sample according to the class label of the most similar available samples. As the K-NN algorithm is performing the classification task after the arrival of a new input sample, it is also called a lazy learner algorithm. Similar to the decision tree classifier, the K-NN algorithm is also used for both classification and regression.

Random Forest: Random forest comes under ensemble classifier. It uses many individual decision trees and then ensembles them together for classifying the input samples. The term bagging is used for the step of training the decision trees. Here the individual decision trees are performing the classification task and assigning the class label to the input sample. These classification results are combined and the class label which is assigned by more number of decision trees is chosen as the final class label by this technique.

Support Vector Machine (SVM): SVM is a classical machine learning algorithm that can be applied for classification, regression, and outlier detection. The main aim of the support vector machine algorithm is to separate the given data samples into two different classes. This is achieved by finding a hyper plane in an N-dimensional space, where N is the number of features used for the classification of the given data samples. For the given data samples, multiple hyper planes are possible, the task is to choose a hyper plane with maximum margin. Margin measures the distance between the data points of the different classes. Data points closer to hyper plane are termed support vectors.

Naïve Bayes Classifier: The collection of classification algorithms working based on the principle of Bayes Theorem are called Naive Bayes classifiers.

The Bayes' Theorem is useful for calculating the likelihood of an event occurring given the probability of a previous event. The following equation expresses the Bayes theorem mathematically:

$$P(A \mid B) = \frac{P(B \mid A)P(A)}{P(B)} \tag{11.1}$$

11.3.1.2 Regression

Regression algorithms are commonly used supervised learning algorithms for predicting the output if it takes some continuous values. Classification is used for predicting the output, if it takes some categorical values like yes/no, correct/incorrect, male/female, etc. Weather prediction, Stock market prediction is the popular examples of scenarios where we can apply regression algorithms for predicting the temperature value or stock price, etc. The different types of available Regression algorithms under supervised learning are as follows:

- Linear Regression

- Non-Linear Regression
- Regression Trees
- Bayesian Linear Regression
- Polynomial Regression

11.4 Machine Learning Algorithm for Solar Tracking System

The proposed solar tracking system uses two supervised learning algorithms namely classification and regression for increasing the energy gain. Binary Classification is utilized to classify the LDR output as "Faulty" or "Non-Faulty". Support Vector Machines (SVMs) classification algorithm is chosen to classify the LDR readings as "Faulty" or "Non-Faulty". If the LDR readings are "Faulty", then the linear regression is applied to predict the angle of rotation and this will be given as input to the servo motor accordingly, the solar panel will get rotated and inclined towards the direction of sun rays.

11.4.1 SVM for solar tracking system

SVM is introduced in the year 1960 and got its full form in the year 1990. SVM is a simple yet efficient supervised machine learning algorithm useful for accomplishing both the classification and regression tasks. SVM is considered a powerful classification algorithm due to its ability to handle both continuous and categorical output variables.

Building the SVM model is a task of representing different classes in a multidimensional space separated by a hyper plane. The main objective of the model is to generate the hyper plane in an iterative phase through the training phase. Generation of hyper planes will appropriately segregate the data into different suitable classes. The accuracy of the model can be further improved by training the model with more samples so that the error can be minimized. The number of hyper planes depends on the number of classes. In the binary classification problem, it is sufficient to generate a single hyper plane to divide the data into two classes.

The various important terminologies associated with the SVM technique are listed as follows:

- Support Vectors – Data points that are near the hyper plane are called support vectors. A separating line will be defined with the help of these data points.

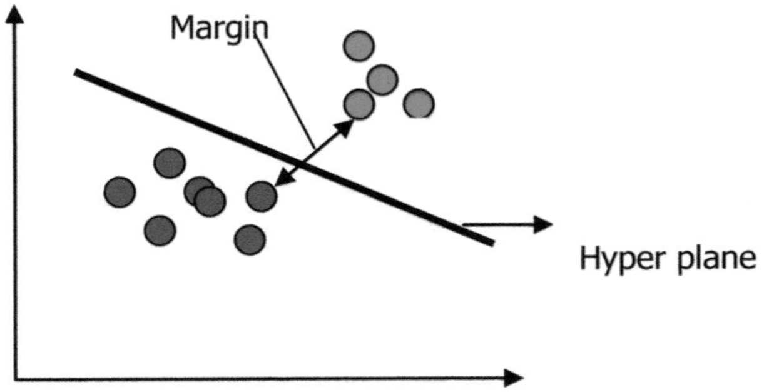

Figure 11.3 SVM classification.

- Hyper plane – Hyper plane is also called a decision plane or space, which divides the input data into different classes.
- Margin – It may be defined as the gap or distance that exists between two lines on the closest data points of different classes. It is calculated by measuring the perpendicular distance between the line and the support vectors. A large margin is always desirable.

11.4.1.1 Steps in python implementation of SVM

- Import the set of needed python libraries like numpy, pyplot,scipy, seaborn, and pandas using import statements.
- Import the data set using the read_csv method available in pandas.
- Split the dataset into training and test samples using the train_test_split function of sklearn.
- Classify the predictors.
- Initialize Support Vector Machine classifier using SVC function of sklearn.
- Fit the training data samples by invoking the fit method.
- Predict the classes for test samples by calling predict method
- Check the predicted results of the test set by calculating the accuracy of the classification using the confusion matrix available in sklearn
- Output:

Accuracy of SVM Classifier for the Given Solar Dataset: 0.934

11.4.2 Linear regression for solar tracking system

Linear regression falls under the supervised learning technique, used to model the relationship between input data samples and the output variable by fitting a linear equation to the given data samples. Initially, the relationship between the different variables in the given data samples is modeled and then fits into a linear model. If there exists a relationship between the different variables, then the linear regression works well for predicting the output values. If there exists no relationship among the different variables, then the model trained using linear regression will become useless, as it cannot able to predict the accurate value of the output variable. The correlation coefficient is a numerical metric, which measures the relationship between two variables. It will take the value between -1 and 1 signifying the relationship strength of the variables. Linear regression model takes the equation of the form $Y = a + bX$, where 'b' represents the slope of the line and 'a' is the intercept.

11.4.2.1 Steps in python implementation of linear regression

- Import the set of needed python libraries like numpy, pyplot, and pandas using import statements.
- Import the data set using the read_csv method available in pandas.
- Split the dataset into training and test samples using the train_test_split function of sklearn.
- Classify the predictors.
- Initialize linear regression object using Linear Regression function of sklearn.
- Train the linear regression model using the training data sets by invoking the fit method
- Make predictions in the testing set by calling the predict method.
- Check the predicted results of the test set by calculating the accuracy of the regression using mean_squared_error and r2_score available in sklearn
- Plot the results (if needed)

11.5 Implementation

// SINGLEAXIS TRACKER with two LDRs connected at A0 and A1 pin (Analog pins) and servo motor is connected to the pin number 9 respectively

```
#include <Servo.h>
//Initialize servo motor
Servo smt;
//Initialize the initial pos to 90
Int init_pos = 90;
//set pin numbers for LDR 1 and LDR2
int L1 = A0;
int L2 = A1;
// Threshold resistance is initialized to 5
int err = 5;
int servo=9;
void setup()
{
smt.attach(servo);
// Pins connected to LDR1 and LDR2 are enabled as
input pins
pinMode(L1, INPUT);
pinMode(L2, INPUT);
//initlaize the position of servo motor
smt.write(init_pos);
delay(2000);
}
void loop()
{
//Read resistance value from LDR1
int R1 = analogRead(L1);
//Read resistance value from LDR2
int R2 = analogRead(L2);
//Use SVM classification to check whether resistance
reported by LDR's are valid
int a = ML(R1,R2);
if (a==1) // resistance reported by LDR's are correct
{
//Find the difference between the resistance
int d1= abs(R1 - R2);
```

```
int d2= abs(R2 - R1);
if((d1 <= error) || (d2 <= error))
{
// Resistance difference less than the threshold,
then do nothing
}
else {
      // According to the resistance, adjust the
position of servo motor towards the sun
      if(R1 > R2){
            init_pos = --init_pos;
      }
      if(R1 < R2){
            init_pos= ++init_pos;
      }
}
smt.write(init_pos);
delay(100);
}
else
{ // Use linear regression algorithm to learn the
position from log records.
learn();
}
}
```

// DUALAXIS TRACKER with four LDRs connected at A0, A1, A2, and A3 pin (Analog pins) and two servo motors are connected in horizontal and vertical directions in pin number 9 and 10 respectively

```
#include <Servo.h>
//Initialize servo motor position (horizontal)
Servo smt_horizon;
int sh = 180;
int shLimitHigh = 175;
int shLimitLow = 5;
//Initialize servo motor position (horizontal)
Servo smt_vert;
int sv = 45;
int svLimitHigh = 60;
```

```
      int svLimitLow = 1;

      int left_ldr = A0;
      int right_ldr = A3;
      int leftdown_ldr =A1;
      int rightdown_ldr = A2;

      void setup(){
      smt_horizon.attach(9);    smt_vert.attach(10);
smt_horizon.write(180); smt_vert.write(45);
      delay(2500);
      }
      void loop(){
      //Read resistance values from four LDRs
      int left = analogRead(left_ldr);
      int right = analogRead(right_ldr);
      intleftdown = analogRead(leftdown_ldr);
      intrightdown = analogRead(rightdown_ldr);
//Use SVM classification to check whether resistance
reported by LDR's are valid
int a = ML(left,right,leftdown,rightdown);
if (a==1) // resistance reported by LDR's are correct
{
      int dtime = 10;
      int tol = 90;
      int averaget = (left + right) / 2;
      int averagd = (leftdown + rightdown) / 2;
      int averageleft = (left + leftdown) / 2;
      int averageright = (right + rightdown) / 2;
      int diff_vert = averaget - averaged;
      int diff_horizon = averageleft - averageright;

      if (-1*tol>diff_vert || diff_vert>tol)
      {
      if (averaget> averaged)
      {
      sv = ++sv;
      if (sv>svLimitHigh)
      {sv = svLimitHigh;}
      }
      else if (averaget< averaged)
```

```
{sv= --sv;
if (sv<svLimitLow)
{ sv =svLimitLow;}
}
vertical.write(sv);
}
if (-1*tol>diff_horizon || diff_horizon>tol)
{
if (averageleft>averageright) //// Need rota-
tion for servo motor connected in vertical direction
{
sh = --sh;
if (sh<shLimitLow)
{
sh = shLimitLow;
}
}
else if (averageleft<averageright) // Need
rotation for servo motor connected in horizontal
direction
{
sh = ++sh;
if (sh>shLimitHigh)
{
sh = shLimitHigh;
}
}
else if (averageleft = =averageright)
{
delay(5000);
}
horizontal.write(sh);
}
delay(dtime);
}
else
{ // Use linear regression algorithm to learn the
position from log records.
learn();
}
}
```

11.6 Conclusion

The suggested dual-axis solar tracker uses little energy and can follow the sun in both directions. Machine-learning-based classification and regression techniques are used to assess the findings of single and dual-axis solar tracking systems. If the classification algorithm detects incorrect results, the linear regression approach is used to anticipate the direction of the solar panel based on past log data, and the servo motor is provided with the necessary input to rotate the panel appropriately. Future studies will look at how to improve prediction using evolutionary algorithms and multi-criteria decision-making strategies. As a result, by enhancing forecast accuracy, the solar tracking system's energy gain may be maximized, and total efficiency can be enhanced.

References

[1] G. E. G. Mustafa, B. A. M. Sidahmed and M. O. Nawari, (2019), The Improvement of LDR Based Solar Tracker's Action using Machine Learning, IEEE Conference on Energy Conversion, Yogyakarta, Indonesia, doi: 10.1109/CENCON47160.2019.8974834.

[2] H. Mohaimin, M. R. Uddin and F. K. Law, (2018), Design and Fabrication of Single-Axis and Dual-Axis Solar Tracking Systems, IEEE Student Conference on Research and Development, Selangor, Malaysia, doi: 10.1109/SCORED.2018.8711044.

[3] S. Ray and A. K. Tripathi, Design and development of Tilted Single Axis and Azimuth-Altitude Dual Axis Solar Tracking systems, IEEE 1st International Conference on Power Electronics, Intelligent Control and Energy Systems, Delhi, doi: 10.1109/ICPEICES.2016.7853190.

[4] G. Boyle. (2004), Renewable Energy: Power for a Sustainable Future, 2nd ed. Oxford, UK: Oxford University Press.

[5] Mr.Solar, (2015), Photovoltaic Effect [Online], vailable: http://www.mrsolar.com/photovoltaic-effect/

[6] Jason Brownlee (2016), Machine Learning Algorithms, https://machinelearningmastery.com/linear-regression-for-machine-learning/

[7] http://www.stat.yale.edu/Courses/1997-98/101/linreg.html

[8] https://analyticsindiamag.com/understanding-the-basics-of-svm-with-example-and-python-implementation/

[9] https://towardsdatascience.com/svm-implementation-from-scratch-python-2db2fc52e5c2

[10] SandeepKhurana (2017), Linear Regression with example, https://towardsdatascience.com/linear-regression-with-example-8daf6205bd49

[11] https://scikit-learn.org/stable/auto_examples/linear_model/plot_ols.html
[12] Savan Patel (2017), Machine Learning 101, https://medium.com/machine-learning-101/chapter-2-svm-support-vector-machine-theory-f0812effc72
[13] M.U.H. Joardder, M.H. Masud, (2017), Solar Pyrolysis, Clean Energy for Sustainable Development.
[14] F. Schenkelberg (2015), Reliability modeling and accelerated life testing for solar power generation systems, Reliability Characterization of Electrical and Electronic Systems.
[15] E. Zarza Moya (2012), Parabolic-trough concentrating solar power (CSP) systems, Concentrating Solar Power Technology.
[16] O. May Tzuc and E. Cruz May (2019), Sensitivity Analysis with Artificial Neural Networks for Operation of Photovoltaic Systems, Artificial Neural Networks for Engineering Applications.
[17] Zhenya Liu (2015), Innovation in Global Energy Interconnection Technologies,Global Energy Interconnection.
[18] Juan Reca-Cardeña and Rafael López-Luque, (2018), Design Principles of Photovoltaic Irrigation Systems, Advances in Renewable Energies and Power Technologies.
[19] JACOB MARSH.(2018), SOLAR TRACKING SYSTEMS, https://news.energysage.com/solar-trackers-everything-need-know/
[20] W.G. Le Roux, J.P. Meyer, (2017), Small-Scale Dish-Mounted Solar Thermal Brayton Cycle, Clean Energy for Sustainable Development.
[21] Thallada Bhaskar, Mukesh Kumar Poddar, (2013), Thermochemical Route for Biohydrogen Production, Biohydrogen.
[22] R. Madavan, Sujatha Balaraman and Saroja S, "Multi Criteria Decision Making Methods for Ranking the Liquid Insulation System Based on its Performance Characteristics with Anti-Oxidants under Accelerated Aging Conditions", IET Gener. Transm. Distrib., Vol., No. 2017.
[23] Rengaraj Madavan, Subbaraj Saroja, "Decision making on the state of transformers based on insulation condition using AHP and TOPSIS methods", IET Sci. Meas. Technol. Vol. 14, No.2, pp.137–145, 2020.
[24] Victor Grigoriev, KyprosMilidonis, Manuel Blanco, (2020), Sun tracking by heliostats with arbitrary orientation of primary and secondary axes,*Solar Energy*,Volume 207, Pages 1384–1389.
[25] L.M. Fernández-Ahumada, J. Ramírez-Faz, R. López-Luque, M. Varo-Martínez, I.M. Moreno-García, F. Casares de la Torre, (2020), Influence of the design variables of photovoltaic plants with two-axis solar tracking on the optimization of the tracking and backtracking trajectory, *Solar Energy,* Volume 208, Pages 89–100.

[26] Manuel G. Satué, Fernando Castaño, Manuel G. Ortega, Francisco R. Rubio, (2020), Auto-calibration method for high concentration sun trackers, *Solar Energy*, Volume 198, Pages 311–323,

[27] Pedro Henrique Alves Veríssimo, Rafael Antunes Campos, Maurício Vivian Guarnieri, João Paulo Alves Veríssimo, Lucas Rafael do Nascimento, Ricardo Rüther, (2020), Area and LCOE considerations in utility-scale, single-axis tracking PV power plant topology optimization, *Solar Energy*, Volume 211, Pages 433–445.

[28] A. Barbón, C. Bayón-Cueli, L. Bayón, P. FortunyAyuso, (2020), Influence of solar tracking error on the performance of a small-scale linear Fresnel reflector, *Renewable Energy*, Volume 162, Pages 43–54.

[29] L.M. Fernández-Ahumada, J. Ramírez-Faz, R. López-Luque, M. Varo-Martínez, I.M. Moreno-García, F. Casares de la Torre, (2020), A novel backtracking approach for two-axis solar PV tracking plants, *Renewable Energy*, Volume 145, Pages 1214–1221.

[30] Jose A. Carballo, Javier Bonilla, Manuel Berenguel, JesúsFernández-Reche, GinésGarcía, (2019),New approach for solar tracking systems based on computer vision, low cost hardware and deep learning, *Renewable Energy*, Volume 133, Pages 1158-1166, ISSN 0960-1481, https://doi.org/10.1016/j.renene.2018.08.101.

[31] Yeguang Hu, Hao Shen, Yingxue Yao, (2018), A novel sun-tracking and target-aiming method to improve the concentration efficiency of solar central receiver systems, *Renewable Energy*, Volume 120, Pages 98–113, ISSN 0960-1481, https://doi.org/10.1016/j.renene.2017.12.035.

[32] NadiaAL-Rousan, NorAshidi MatIsa, MohdKhairunaz MatDesa, (2018), Advances in solar photovoltaic tracking systems: A review, *Renewable and Sustainable Energy Reviews*, Volume 82, Part 3, Pages 2548–2569, ISSN 1364-0321, https://doi.org/10.1016/j.rser.2017.09.077.

[33] VijayanSumathi, R. Jayapragash, AbhinavBakshi, Praveen Kumar Akella, (2017), Solar tracking methods to maximize PV system output – A review of the methods adopted in recent decade, *Renewable and Sustainable Energy Reviews*, Volume 74, Pages 130-138, ISSN 1364–0321, https://doi.org/10.1016/j.rser.2017.02.013.

[34] R.G. Vieira, F.K.O.M.V. Guerra, M.R.B.G. Vale, M.M. Araújo, 2016, Comparative performance analysis between static solar panels and single-axis tracking system on a hot climate region near to the equator, *Renewable and Sustainable Energy Reviews*, Volume 64, Pages 672–681, ISSN 1364-0321, https://doi.org/10.1016/j.rser.2016.06.089.

[35] HosseinMousazadeh, AlirezaKeyhani, ArzhangJavadi, HosseinMobli, Karen Abrinia, Ahmad Sharifi, (2009), A review of principle and

sun-tracking methods for maximizing solar systems output, *Renewable and Sustainable Energy Reviews*, Volume 13, Issue 8, Pages 1800–1818, ISSN 1364-0321, https://doi.org/10.1016/j.rser.2009.01.022.

[36] Puneet Joshi, Sudha Arora, (2017), Maximum power point tracking methodologies for solar PV systems – A review, *Renewable and Sustainable Energy Reviews*, Volume 70, Pages 1154–1177, ISSN 1364-0321, https://doi.org/10.1016/j.rser.2016.12.019.

[37] G. Dileep, S.N. Singh, (2015), Maximum power point tracking of solar photovoltaic system using modified perturbation and observation method, R*enewable and Sustainable Energy Reviews,* Volume 50, Pages 109–129, ISSN 1364-0321, https://doi.org/10.1016/j.rser.2015.04.072.

[38] Saban Yilmaz, Hasan RizaOzcalik, Osman Dogmus, FurkanDincer, OguzhanAkgol, MuharremKaraaslan, (2015), Design of two axes sun tracking controller with analytically solar radiation calculations, *Renewable and Sustainable Energy Reviews*, Volume 43, Pages 997–1005, ISSN 1364-0321, https://doi.org/10.1016/j.rser.2014.11.090.

[39] Ravinder Kumar Kharb, S.L. Shimi, S. Chatterji, Md. Fahim Ansari, (2014), Modeling of solar PV module and maximum power point tracking using ANFIS, R*enewable and Sustainable Energy Reviews*, Volume 33, Pages 602–612, ISSN 1364-0321, https://doi.org/10.1016/j.rser.2014.02.014.

Chapter 12

Gain and Bandwidth Enhancement of Pentagon Shaped Dual Layer Parasitic Microstrip Patch Antenna for WLAN Applications

Ambavaram Pratap Reddy[1], and Pachiyaannan Muthusamy[2]

[1,2]Advanced RF Microwave & Wireless Communication Laboratory, Vignan's Foundation for Science Technology and Research (Deemed to be University), Vadlamudi, Andhra Pradesh, India
Email: pratap.phd5001@gmail.com; pachiphd@gmail.com

Abstract

A new Pentagon two-layer patch antenna has been designed, featuring four parasitic patches and proximity coupling feeding. Initially, a single-layer antenna was suggested with an FR-4 substrate height of I.6mm. Based on the observation that a dual-layer antenna with the same substrate material and the same dielectric material height of 1.6 mm increases all parameters with excellent values was offered to improve the results. Two parasitic elements and four parasitic patches make up the dual layer of the Pentagon-shaped patch antenna. The antenna gain may be enhanced from –40dB to 44dB with a –44.34 dB reflection coefficient and a gain of roughly 6.02dB to 7.09 dB, according to the data. The suggested antenna will work at a frequency of 5.4 GHz. The verification of gain increase was examined using a parasitic analysis of the patches. The suggested antenna has a total length of 30mm and a width of 40mm. CSTMW 2018 was used to simulate the suggested design's outcomes. A vector network analyzer was used to construct and evaluate the parasitic patch antenna that was proposed. Both measured and simulated findings are quite accurate and should be used in WLAN applications. The suggested antenna is ideally suited for WLAN applications because of these features.

12.1 Introduction

In today's wireless communication development and growth of their applications, printed slot antennas are now being considered for use in wideband communication systems due to their appealing features, such as wide impedance bandwidth, compact size, simple structure, low cost, and easy integration with monolithic microwave integrated circuits. Printed slot antennas are now being considered for use in wideband communication systems due to their appealing features. A number of printed slot antenna designs have been described for wideband applications, including squares [1, 2] and rectangles. Impedance bandwidths of between 60% and 104% have been reported in these studies. Multiple parasitic patches and shorting vias have been described in [3] for the purpose of enhancing the bandwidth of microstrip patch antennas. In [4], the author designed a multiple beam parasitic array radiator (MBPAR) antenna for obtaining good return loss and radiation patterns. A printed microstrip-line-fed slot antenna with a pair of bandwidth-enhancing parasitic patches is recommended in [5]. To improve the results, a c- shaped parasitic element was put around the microstrip patch [6]. A wideband parasitic element was employed as a decoupling structure to improve isolation for UWB applications in [7]. [8] Designed a dual-polarized multilayer patch antenna with layered parasitic patches and CSRRs to increase bandwidth. The author presented a single-layer roger substrate with a square radiating patch and two parasitic strip lines in[9] PIN diodes are put between the patch and the parasitic strip line for higher gain and cross-polarization isolation. In [10], a multi-layer dual-polarized antenna operating in the Ku band was designed for low return loss and high bandwidth. In [11], an analysis was carried out for multilayer microstrip patch antennas with different heights and different shapes.

In [12], a two-layer single-feed, compact, and wideband microstrip antenna was developed with the purpose of increasing gain and efficiency. The design of a multilayer microstrip antenna in which metamaterial is placed on the substrate to boost gain is described in [13]. For WLAN and WiMAX, [14] designed a stacked triple-band meta-mode antenna with a coaxial feed and two symmetrical comb shaped split ring resonators (CSSRR). A stacked patch antenna with a metamaterial has been used in [15] to obtain significant gain, directivity, and low return loss for broadband applications. An H-shaped microstrip antenna was constructed to improve gain, return loss, and bandwidth at a frequency of 4.8GHz [16]. A microstrip antenna with coaxial probe feeding technology was invented [17] to make the antenna better for WiMAX and WLAN use.

In this work, a dual-layer pentagon antenna has been proposed for the improvement of the gain and bandwidth. The pentagon antenna had some analysis in order to improve the gain and bandwidth enhancement. Initially, a single-layer pentagon is proposed, followed by a dual-layer with two parasitic elements, and finally, a four-element parasitic is offered for further development. According to the observations, the results are better when there is a double layer of four parasitic elements.

12.2 Pentagon Single Layer Design

The desired pentagon single-layer antenna with a radius of R1 10mm on an FR4 substrate and a height of 1.6mm is shown in Figure 12.1. Quarter wave feeding is used to provide impedance matching. The proposed antenna's reflection coefficient is –23dB with a VSWR of 2, as shown in Figure 12.2. The suggested antenna has a gain of 4.98dB, as shown in Figure 12.3. The operational bandwidth ranges from 5.60 to 5.79 GHz. Table 12.1 displays the dimensions of the proposed antenna.

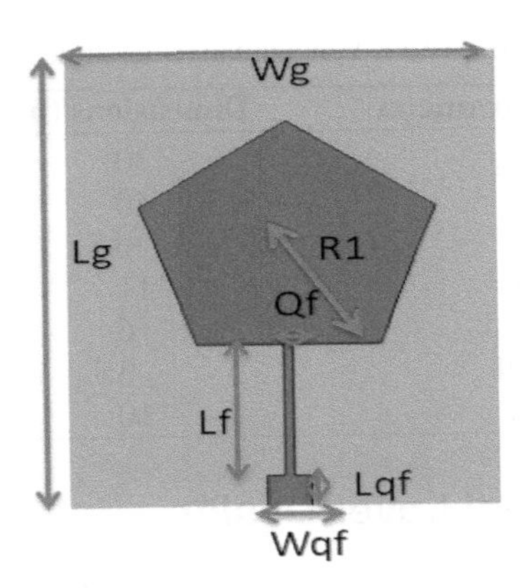

Figure 12.1 Single layer pentagon Antenna.

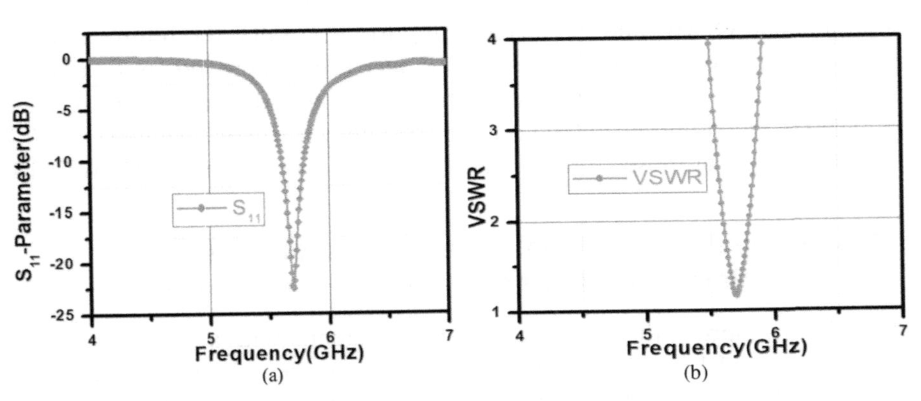

Figure 12.2 Proposed results (a) S_{11}-Parameter (b) VSWR.

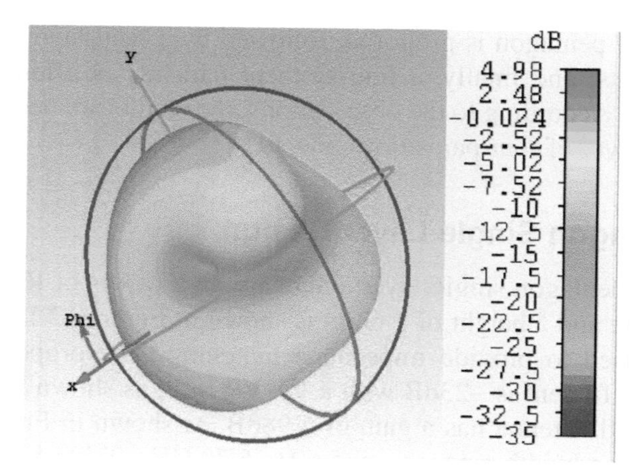

Figure 12.3 Proposed antennas Gain at 5.3GHz.

Table 12.1 Pentagon antenna dimensions.

Parameter	Dimensions (mm)
Wg	30
Lg	40
Wf	0.7
Lf	11
Wqf	6
Lqf	3.1
R1	10

12.3 Pentagon Dual Layer Design

Figure 12.4 depicts a dual-layer suggested antenna with the same FR4 height of 1.6mm for improved gain. Figure 12.5 depicts the proposed antenna's s-parameter, VSAW, and gain. According to the results, the return loss has been minimized by −10dB, the VSWR has been improved to 1, and the gain has increased by 1dB. The operational bandwidth ranges from 5.2–5.54 MHz The proposed antenna center frequency has been changed from 5.7 to 5.4 MHz the proposed design results were simulated using CSTMW 2018.

12.4 Analysis of the Dual Layer Pentagon with Two Parasitic Elements

In order to improve the gain, the dual-layer antenna with a rhombus-shaped parasitic patch P1 and P2 were positioned on both sides of the feed line shown in Figure 12.6. Figure 12.7 illustrates its low reflection coefficient with gain compared with the single layer antenna.

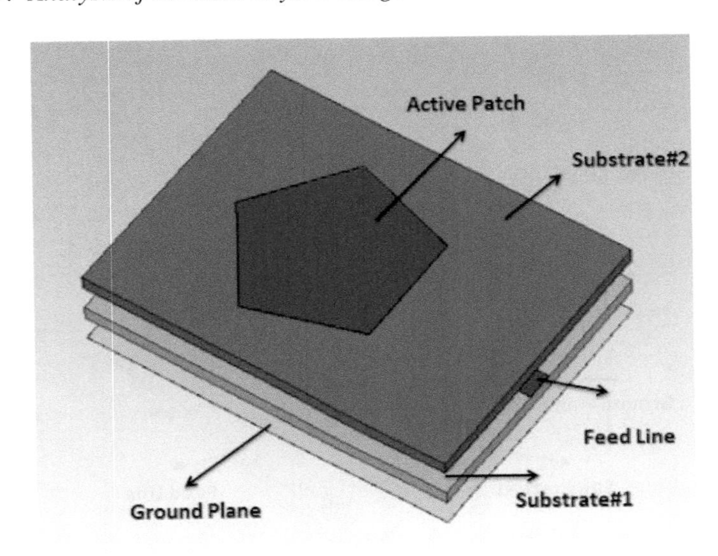

Figure 12.4 Proposed dual layer Antenna.

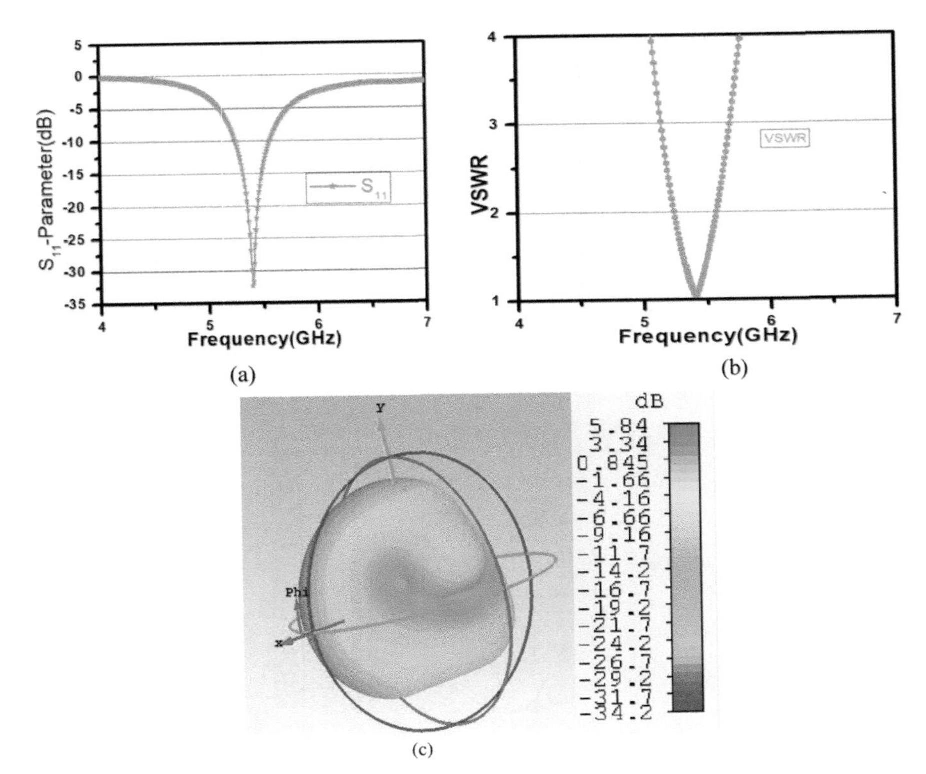

Figure 12.5 Proposed antenna results (a) S_{11}-Parameter (b) VSWR (c) Gain.

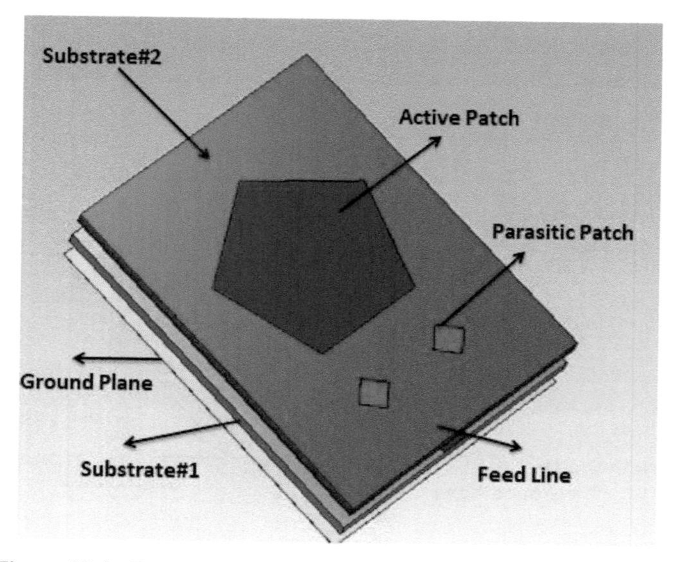

Figure 12.6 Proposed antenna dual layers with two parasitic element.

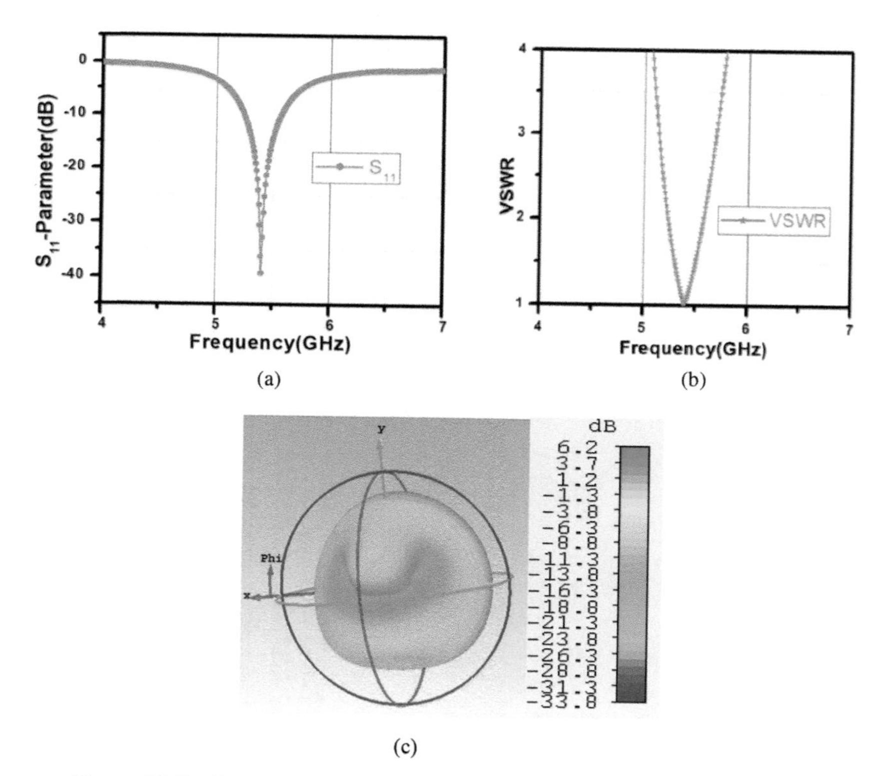

Figure 12.7 Proposed antenna results (a) S_{11}-Parameter (b) VSWR (c) Gain.

12.5 Analysis of the Dual Layer Pentagon with Two Parasitic Elements

For more and more improvement, two more parasitic components, P3 and P4, are positioned in between the proposed antenna, as shown in Figure 12.8. The enhanced required results of the proposed antenna are shown in Figure 12.9. Given the enhanced bandwidth of 5.22–5.98MHz, the four-element parasitic design improves gain by 1.1dB while maintaining a very low return loss of –44dB at the target frequency, compared to the two-element parasitic design. As shown in Figure 12.10, the surface current distribution of the antenna dual-layer antenna is analyzed to understand the effect caused by the four parasitic patches. The strong current distribution at 5.4 GHz is mainly concentrated on the pentagon dual layer patch as well as the parasitic components, as seen in Figure 12.10. Figure 12.11 shows the quasi Omni directional radiation pattern at 5.4GHz.With a canter frequency of 5.4GHz, the low return loss is reached by –44dB with a considerable increase in gain and bandwidth.

The study was carried out in order to enhance the gain, return, loss, and bandwidth of the proposed design. The single layer penton antenna is discussed in Section 12.3 along with the findings. Based on the observations, gain enhancement is needed for practical usage. Section 12.4 discusses a dual-layer antenna with the same substrate and 1.6mm height that improves a single-layer antenna. Two parasitic elements are positioned for further gain enhancement. It needs to be noted that this improves the gain and loss described in the Section 12.5. By positioning the extra two elements between the planned

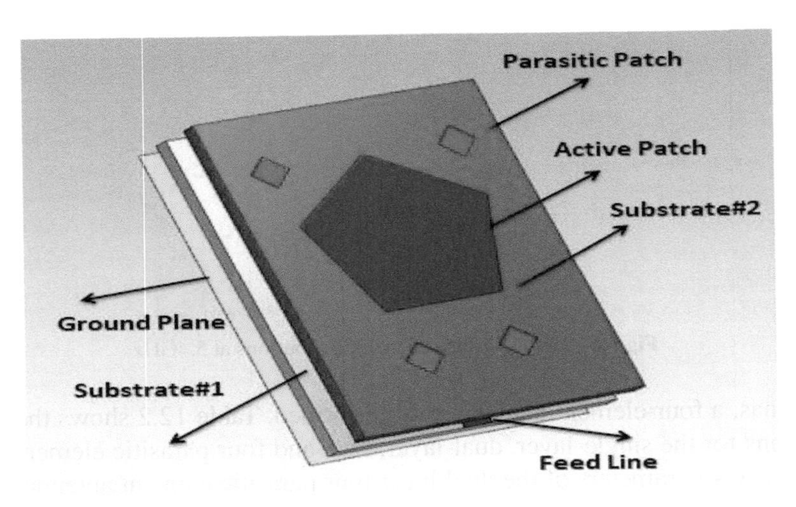

Figure 12.8 Proposed antenna dual layers with four parasitic elements.

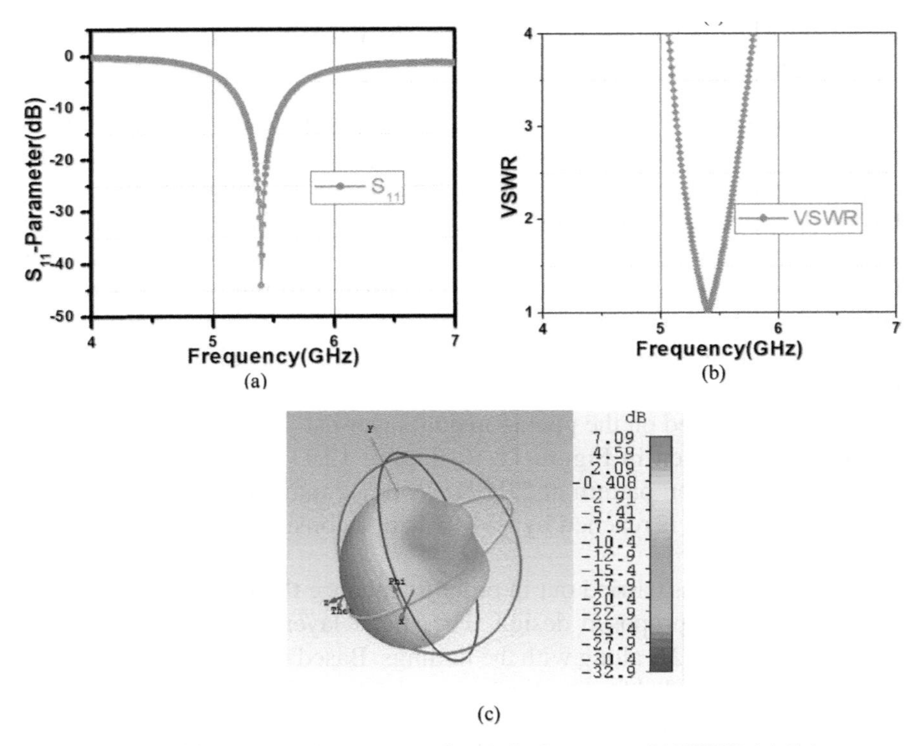

Figure 12.9 Proposed antenna results (a) S_{11}-Parameter (b) VSWR (c) Gain.

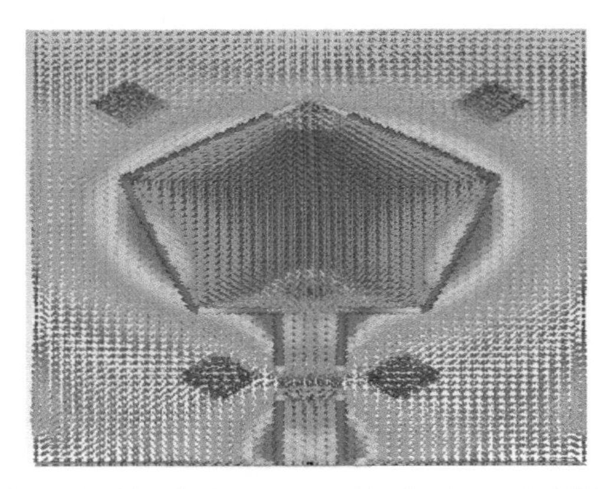

Figure 12.10 Surface current distributions at 5.4GHz.

antennas, a four-element analysis was performed. Table 12.2 shows the comparisons for the single-layer, dual-layer, two, and four parasitic element analyses. The s-parameters of the dual layer four parasitic element antennas were simulated and measured in Figure 12.11. The suggested antenna prototype and

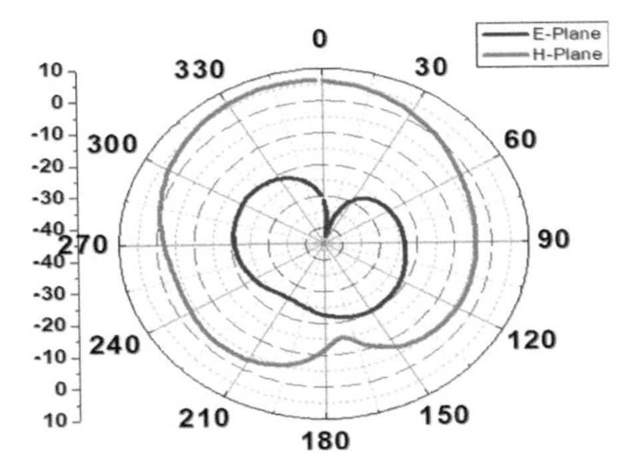

Figure 12.11 Radiation pattern at 5.4GHz results and discussion.

Table 12.2 Comparison analyses.

S.no	Single Layer	Dual Layer	Two parasitic elements	Four Parasitic Elements
S11	–23dB	–33dB	–40dB	–44dB
VSWR	1.2	1	1	1
Gain(dB)	4.98	5.84	6.2	7.09
Bandwidth	5.60–5.79(MHz)	5.24–5.54(MHz)	5.25–5.75(MHz)	5.22–5.98(MHz)

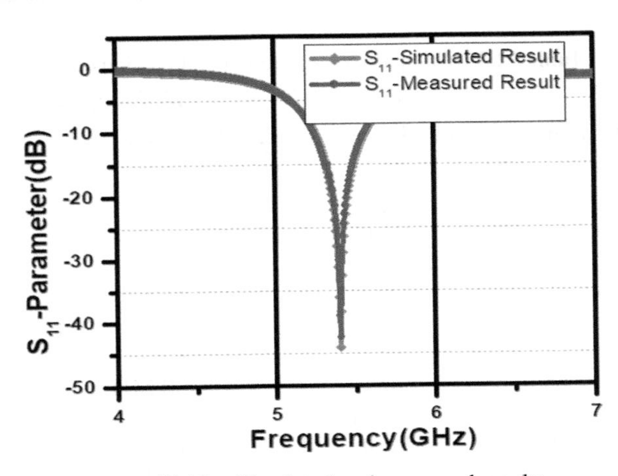

Figure 12.12 Simulated and measured results.

measurement setup are shown in Figure 12.12. Table 12.3 illustrates the comparison between this current study and previously published studies.

12.6 Conclusion

In this work, a new pentagon microstrip patch antenna with four parasitic patches has been developed for WLAN applications to improve gain

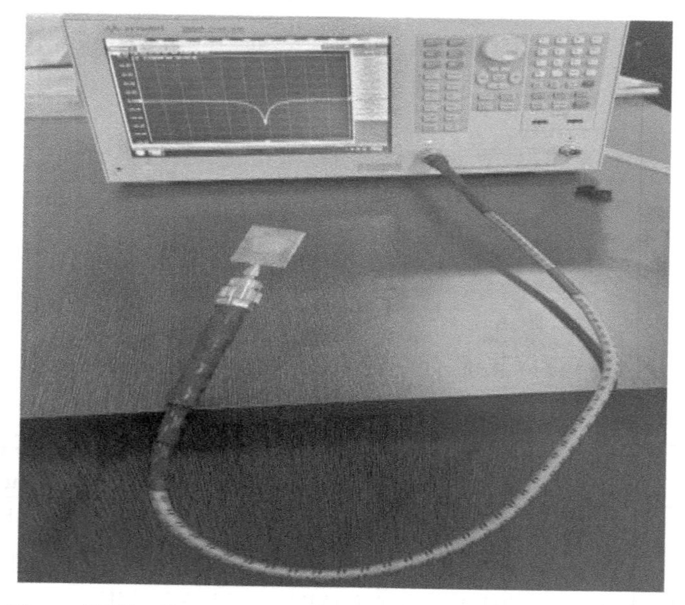

Figure 12.13 Prototype proposed antenna with measurement setup.

Table 12.3 Comparisons of present and previous works.

Reference Antenna	Dimensions(mm)	No.of Layers	Gain(dB)	Bandwidth (GHz)	Frequency (GHz)
16	28. mm × 24. mm²	Single Layer	6.46	–	4.8 GHZ
17	90 mm × 90 mm²	Single Layer	6.6,5.5,6	–	3.47 GHz, 3.71 GHz
3	36 × 39 mm²	Single Layer	3	5.46 to 6.27 and 5.5 to 6.55	5.7 GHz, 6 GHz
6	65 × 65 mm²	Single Layer	4	30 MHz	1.6 GHz
10	40 × 30 mm²	Multi Layers	12.5	14.5 to 18.5 GHz	16 GHz
11	151 ×151 mm²	Dual Layers	4.64	2.333 to 2.377 GHz	2.35 GHz
Proposed Work	30 × 40 mm²	Dual Layers	7.09	5.22 to 5.98 MHz	5.4 GHz

improvement. Improved and return loss may be achieved when parasitic patches are included in the construction of the proposed antenna, and thus the bandwidth can be effectively extended. The proposed antenna is designed to have an S11of -44dB, a gain of 7.02dB, and a bandwidth of 5.22-5.98MHz at the center frequency of 5.4GHz. The impacts of parasitic factors on the antenna's performance have been investigated using a parametric study.

References

[1] W. S. Chen, and F. M. Hsieh, "A broadband design for a printed isosceles triangular slot antenna for wireless communications," Microw. J., vol. 48, no. 7, pp. 98–112, Jul. 2005.

[2] W. S. Chen, "A novel broadband design of a printed rectangular slot antenna for wireless communications," Microw. J., vol. 49, no. 1, pp. 122–130, Jul. 2006.

[3] Kai Da Xu,HanXu,YanhuiLiu,JianxingLi,and Qing Huo Liu, "Microstrip Patch Antennas with Multiple Parasitic Patches and Shorting Vias For Bandwidth Enhancement," published in IEEE Access , DOI :10.1109/ ACCESS.2018.2794962.(2017).

[4] Qiuyan Liang, Baohua Sun, and Gaonan Zhou "Multiple Beam Parasitic Array Radiator Antenna for 2.4-GHz WLAN Applications,": DOI 10.1109/LAWP.2018.2880208, IEEE Antennas and Wireless Propagation Letters.(2018).

[5] S.T. Fan, Y.Z. Yin, B. Lee, W Hu, and X. Yang "Bandwidth Enhancement of a Printed Slot Antenna with a Pair of Parasitic Patches,"published in IEEE 2011.

[6] Zhao-Lin Zhang, Kun Wei ,JianXie , Jian-Ying Li, and Ling Wang "The New C-Shaped Parasitic Strip for the Single-Feed Circularly Polarized (CP) Microstrip Antenna Design," International Journal of AntennasandPropagationVolume2019,https://doi.org/10.1155/2019/ 5427595.

[7] Kunming Zhang,Jinkai Li, HuiqingZhai , Zhigang Wang and Yi Zeng "UWB-MIMO antenna decoupling based on a wideband parasitic unit structure," published in IEEE Xplore 2020.

[8] Li Guo, Ming-Chun Tang "A Low-Profile Dual-Polarized Patch Antenna with Bandwidth Enhanced by Stacked Parasitic Elements," published in ICMMIT in 2018 IEEE.

[9] M.Saravanan, A.Priya, "A Compact Frequency and Polarization Reconfigurable Square Patch Antenna for Wireless Communication," published in IEEE 2018.

[10] Yao Zong, Jun Ding, ChenjiangGuo and Jun Zhang "An Improved Broadband Multi- layer Dual-polarized Antenna for UAV Radars," Published in 2019,IEEE.

[11] S.Phani varaprasad & R. Prasad rao "Design and analysis of a multi layer substrate single patch microstrip patch antenna for enhancing the beam width with control on directivity," published in Jan 2017 international Academy of science, Engineering and technology.

[12] Munish Kumar and VandanaNath "Improved Cross Polarization and Wideband Multilayer Wide-slot Microstrip Antenna with Rotated Parasitic Patch," Published in 2017,IEEE.

[13] Shobhit K. Patel, Karan H. Shah and Y.P.Kosta "Multilayer liquid metamaterialradome design for performance enhancement of microstrip patch antenna," February 2018 Microwave and Optical Technology Letters, DOI:10.1002/mop.31024.

[14] Rajiv Mohan David , Mohammad Saadh Aw ,Tanweer Ali, Pradeep Kumar "A Multiband Antenna Stacked with Novel Metamaterial SCSRR and CSSRR for WiMAX/WLAN Applications," published in National Library of Medicine in 2021.

[15] DivyaC ,Koushick "Design and implementation of slotted metamaterial stacked microstrip patch antenna for broadband applications," Published in ICE4CT 2019.

[16] AbhishekMadankar,VijayChakole, SachinKhade "H-slot Microstrip Patch Antenna for 5G WLAN Application," published in IEEE92020; ISBN: 978-1-7281-7089-3.

[17] Prahlad, Radha Anil Kandakatla, PrasannaM,AbhinandanAjitJugale, Mohammed Riyaz Ahmed "Microstrip Patch Slot Antenna Design for WiMAX and WLAN Applications," published in Third International Conference on Trends in Electronics and Informatics (ICOEI 2019) IEEE EXplore Part Number: CFP19J32-ART; ISBN: 978-1-5386-9439-8.

Chapter 13

Quantum Cascade Lasers – Device Modelling and Applications

M. Ramkumar[1], Pon Bharathi[2], A. Nandhakumar[3], P. Srinivasan[4], D. Vedha Vinodha[5], and P. Ashok[6]

[1]Associate Professor, Department of ECE, Sri Krishna College of Engineering and Technology, Coimbatore
[2]Assistant Professor, Department of ECE, Amrita College of Engineering and Technology, Nagercoil
[3]Department of ECE, Dhaanish Ahmed Institute of Technology, Coimbatore
[4]Department of ECE, Amrita College of Engineering and Technology, Nagercoil
[5]Department of ECE, JCT college of Engineering and Technology, Coimbatore
[6]Assistant Professor, Department of ECE, Sri Venkateswara College of Engineering, Chennai, India.
Email: [1]mramkumar0906@gmail.com, [2]bharathpon@gmail.com, [3]nandhakumar3107@gmail.com, [4]srinivasanpaul@gmail.com, [5]dvedha1975@gmail.com; [6]ashokp2k4@gmail.com

13.1 Introduction

Terahertz quantum cascade-lasers (THz QCLs) operating in pulsed mode and at higher temperatures produce larger optical power. Predicting device characteristics in pulsed mode becomes a nontrivial activity, involving the device parameters. They also produce complicated responses in terms of optical output power and emission frequency. In some applications, it is critical to forecast and manage the parametric behaviors, which necessitates creating a relationship between current drive, emission frequency, and optical output power. THz QCLs are potential radiation sources for sub-MHz or even kHz spectral resolution high-resolution spectroscopy. They have an extremely narrow inherent linewidth of roughly 90 Hz. THz QCLs are primarily used in two spectroscopic methods. Absorption spectroscopy is one approach in which the emission of a coherent THz source with small linewidth is frequency scanned through the spectrum absorption feature and the transmission is determined as a function of frequency. The other approach that uses THz QCLs is heterodyne spectroscopy, which is used for distant sensing in astronomy and planetary exploration. Its main advantage over other spectroscopic techniques like Fourier transform and grating spectroscopy is its extraordinarily high spectral resolution. This enables atomic and molecule emission or absorption lines to be resolved spectroscopically. This chapter provides an exhaustive report on device modeling and applications of QCL from the recently available articles. It is compiled and presented in a way to mainly benefits researchers and industry personnel.

13.2 Non-Linear Frequency Generation

In resonantly stimulated nanostructures, second-order optical nonlinearities can be considerably amplified. As a result of massive destructive interference effects, the second-order nonlinear susceptibility can vary by order of magnitude [1]. THz QCLs are used to simulate nonlinearities between subbands. These massive interferences are likewise thought to be the result of second-order nonlinear contributions from several lights and heavy hole states. This framework may be used to create the resonant optical characteristics of nanostructures using a unique, sensitive method to reveal the band structure features of complicated materials.

As a result, the emission of QCLs is suitable for probing low energy excitations. The nonlinear response of the QCL is used. The results clearly reveal that these effects are caused by significant susceptibility interferences between the various nonlinear contributions from the numerous light (LH) and heavy (HH) hole states. The nonlinear conversion may be utilized to examine the complicated QCL band structure in order to obtain information on wavefunction alignment.

13.3 QCL Based Interferometry

Laser Feedback Interferometry (LFI) is one of the numerous materials analysis and imaging techniques that are now commonly utilized at THz frequencies. The laser serves as both a source and a detector. Figure 13.1 depicts the basic LFI arrangement. The term “swept-frequency LFI” refers to a type of LFI sensing. Sweeping the emission frequency of a laser creates an interferogram from which target information such as location, complex reflectivity, and refractive index may be obtained. With bigger frequency shifts, improved

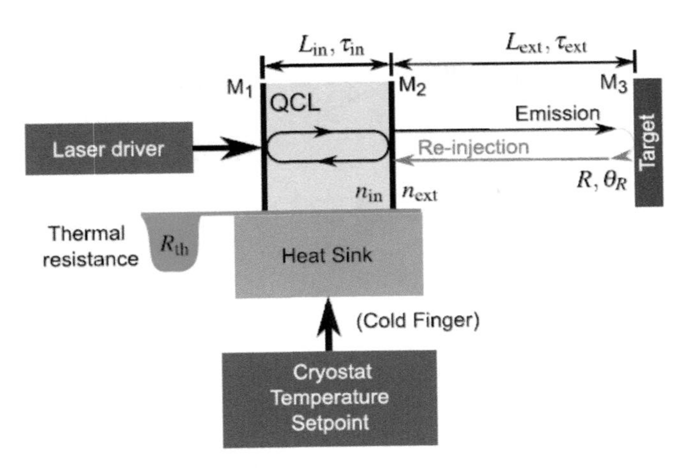

Figure 13.1 Model of a pulsed QCL under optical feedback [2].

LFI results may be produced, making strategies for increasing the range of a laser's frequency sweep desirable.

The QCL employed in this experiment is a 2.59 THz single-mode laser made of GaAs/GaAlAs multilayers, processed into a surface plasmon Fabry-Pérot ridge waveguide, and indium-mounted onto a copper sub-mount. The gadget can operate in continuous wave mode up to 50K cold finger temperatures and generates a maximum of 4mW of optical power from each facet. Aside from the well-known benefits of pulsing (increased optical output power and operating temperature), the form of the driving pulse may be tailored to improve and linearize the range of a QCL's frequency sweep by a precise combination of adiabatic and thermal modulation. This is especially relevant in THz materials investigation, where swept-frequency LFI is used, but it is equally important in any pulsed application where emission frequency changes. It is believed that this approach might be beneficial in additional laser applications requiring a greatly prolonged and linear emission frequency sweeps, such as trace gas detection, imaging, heterodyne mixing, and spectroscopy.

13.4 Frequency Instabilities in THz QCLs

The effect of optical feedback on the frequency stability of THz QCLs [3] is studied in depth using high-resolution heterodyne spectroscopy. The emission from two pairs of QCLs operating at 3.4 or 4.7 THz is mixed with a Schottky diode, and the difference frequency in the GHz region is recorded. The great spectral resolution of less than 1MHz and the temporal resolution

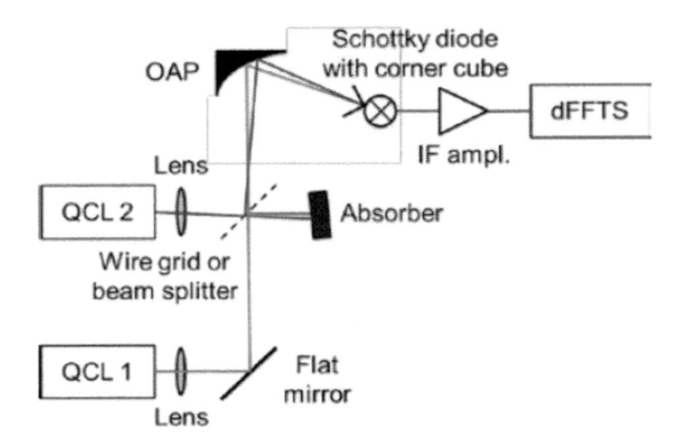

Figure 13.2 Experimental setup for the 4.7-THz experiments [3].

of 1ms allow us to examine feedback-induced frequency instabilities on spectroscopic scales. Frequency shifts of up to 70 MHz have been reported with visual feedback of just $3*10^{-5}$ of the QCL's output power. The experimental setup is shown in Figure 13.2.

The effect is comparable for lasers with Fabry-Pérot resonators and lasers with distributed-feedback resonators. When the feedback is out of phase, mode hopping between two modes of the external resonator happens on a time scale of less than 1ms. This occurs when the QCL's wavelength is adjusted to the desired frequency and mechanical vibrations affect the length of the external resonator. The findings emphasize the critical relevance of suppressing standing waves and external resonator modes in THz spectroscopy with high accuracy and spectral resolution.

13.5 Design Optimization of Cavity in QCLs

The manufacture and characterization of mid-infrared AlInAs/InGaAs/InP QCLs taper quantum cascade lasers are fully described. The tapered shape of the resonator is used to boost output power while maintaining acceptable beam quality. The output beam characteristics of such devices emitting at 4.5m with varied forms (linear, concave, and convex) and angles of the taper waveguide section are optimized. Taper designs enable increased output power while retaining essential TEM00 mode performance. The trade-off between beam quality and output power is always there. Tapered waveguides also minimize optical intensity at the output facet and the consequences of self-heating [4]. Taper geometry lasers are divided into two sections: a narrow ridge waveguide segment that only allows the TEM00 mode to propagate and a tapered region where the mode is adiabatically enlarged, allowing for high optical output power. According to the experimental results, a convex geometry taper laser with a 1.7° taper has the greatest performance, the maximum output power up to 5W, and the minimum horizontal beam divergence of 5.6° in the fundamental mode. The maximum brightness is also a feature of this design.

13.6 THz QCLs Based on HgCdTe Material Systems

The 2.3 THz QCL is designed and manufactured using an active module based on 4 Quantum Wells $GaAs/Al_{0.15}Ga_{0.85}As$ [5]. The constructed THz QCL's P-I-V properties, emission spectra, and far-field distribution of emission intensity are measured. The balancing equation approach is used to simulate the laser properties to assess the possibility of the $Hg_{1-x}Cd_xTe$ structures as a gain medium for QCL operating at a frequency exceeding 6 THz. The creation of Electric Field Domains (EFDs) leads to a drop in output intensity as shown by the emission spectra at different currents. The theoretical model proposed predicts the production of EFDs in the NDR area and qualitatively explains the experimental data. In addition, the far-field distribution of 2.3 THz QCL emission intensity is examined.

To increase the performance and prolong the operating frequency of THz QCLs, it is necessary to prevent the production of EFDs and to apply innovative material systems for lasing at frequencies greater than 6 THz. The analysis of HgCdTe QCLs based on three- and two-well designs using a resonant-phonon depopulation strategy is performed. The improved two-well design of HgCdTe QCL with peak gain reaching 100 cm^{-1} at 200 K demonstrates the highest temperature resistance.

13.7 Optical Beam Characteristics of QCL

The effect of front and rear laser mirror ion cleaning on quantum cascade laser properties is thoroughly investigated [6]. Cleaning the front mirror improves beam symmetry and boosts external differential quantum efficiency, whereas cleaning the rear mirror improves electrical resistance, raises threshold current and has little effect on beam optical polarization. Focused ion beam (FIB) technology has several uses in semiconductor devices, as well as post-fabrication modification of photonic devices to increase performance. The effect of FIB mirror polishing on the beam characteristics of a quantum cascade laser is thoroughly investigated.

The laser under study is a strain-compensated $In_{0.67}Ga_{0.33}As$/$In_{0.36}Al_{0.64}As$/InP quantum cascade laser that has been wet etched into a double trench mesa. Laser mirrors were first cleaved and then machined using a FIB. Ion milling was carried out in the Helios Nano Lab 600 Dual Beam system using a concentrated Ga+ ion beam at 30kV accelerating voltage. The threshold current remained the same after front mirror milling, whereas the external differential quantum efficiency rose. The threshold current rose with the same efficiency after rear mirror milling. Optical polarization was investigated at spatial points with the greatest optical power. Overall, milling the front mirror has a definite favorable influence on quantum cascade laser properties, but milling the back mirror has a detrimental effect.

13.8 Temperature Degradation in THz QCLs

Low operating temperatures are now one of the most serious issues with THz QCLs. The maximum operating temperature for THz QCLs with frequencies close to 3THz is 117K in CW mode and 199.5K in pulsed mode. Moving from 3 to 1THz and 3 to 5THz, however, drastically reduces the output powers and operating temperatures of the THz QCLs due to numerous physical restrictions. Two THz QCL designs [7] are presented based on three- and four-well active modules with gain maxima at 3.2 and 2.3THz, respectively. The laser structures were created using molecular beam epitaxy with an active area thickness of 10m. The Au-Au double metal waveguide is used to make the ridge structures. The device was wire bonded to ridge structures after being dying bonded to a copper sub-mount. The measurements were taken with a laser facet silicon hyper hemispherical lens that was abutted to a laser facet silicon hyper hemispherical lens. The experimental relationship between output power and operating temperature for 3.2THz QCL with a maximum operating temperature of roughly 86 K. There are three equidistant spectral lines corresponding to the longitudinal Fabry-Perot modes with a frequency spacing of 50 GHz for L = 1 mm in observed at 58 K spectrums of 3.2 THz QCL and 2.3 THz QCL.

13.9 Impedance Characteristics of QCLs

Until a full numerical study of the device was provided in 2019 [8] it was not possible to compute the inherent impedance of QCLs. The QCL under consideration for the study works at 9m and has 48 phases. For the first time, the fluctuation of device impedance with respect to optical confinement factor, mirror reflectivity, and spontaneous emission factor is presented. The gadget has an electrical bandwidth of 250GHz and an inherent impedance of 2.07mΩ under normal working circumstances. According to the findings of the investigation, the highest value of impedance is always attained with the least amount of driving current. Furthermore, with a mirror reflectivity (R) of 0.45, the magnitude of the maximum value of the impedance (2.37mΩ) is reached. Furthermore, when the optical confinement factor (Γ) is 0.45 and the drive current is 3A, the maximum bandwidth of 335GHz is obtained. The device also has a flat impedance response with a value of roughly 1m over a large frequency range of 1 to 20GHz, allowing it to be used in 5G wireless networks. QCL's impedance characteristics will help in the construction of correct matching circuits for decreasing reflections and maximizing efficiency.

13.10 Free Space Optical Communication Using QCLs

The performance of a free space optical (FSO) communication connection is investigated in this research by adding device factors that impact the optical output power of the QCLs. The FSO link's transmitter is made up of gain-switched QCL running at 9μm. Short optical pulses with a minimal full width half maximum (FWHM) and peak power are broadcast into the channel as a result of ON-OFF keying. The pulses are attenuated by medium bound losses before arriving at the receiver, which uses a quantum well-infrared photodetector operating at the same wavelength. The block diagram of a generic FSO link is depicted in Figure 13.3.

For the FSO link, the average signal-to-noise ratio (SNR_{avg}), bit error rate (BER), and channel capacity are calculated for all changes in the device parameters, namely optical confinement factor (Γ), spontaneous emission factor (β), and mirror reflectivity (R). The device parameter combination that provides the best connection performance is evaluated. According to the

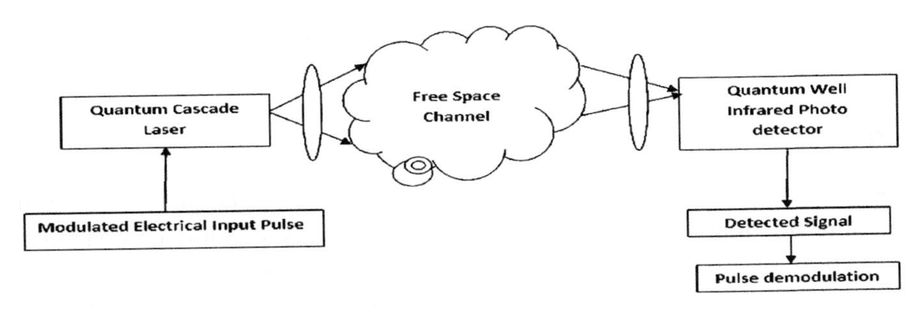

Figure 13.3 Block diagram of a generic FSO Link.

results of the investigation, QCL with $\Gamma = 0.32$ gives the best average SNR of 4.04dB, the lowest BER of 13.02 $*10^{-2}$, and the largest capacity of 1.82 bps/Hz under minimal FWHM conditions. When peak power is limited, QCL with mirror reflectivity $\Gamma = 0.45$ delivers the greatest average SNR of 24.78 dB, the lowest BER of $7.75*10^{-35}$, and the highest peak capacity of 8.24 bps/Hz [9]. Some QCL applications necessita-te the development of pulses with a shorter turn-on latency, while others necessitate the generation of short optical pulses with a low FWHM and peak power. Based on the study results, the optimum feasible combination of operating conditions for the QCL for a certain application may be selected.

13.11 Free Space Optical Communication Using Temperature Dependent QCLs

Temperature-dependent terahertz frequency quantum cascade lasers (QCLs) [10] have the potential to be used in a wide range of innovative applications. Reduced rate equations are used to gain a better knowledge of their behavior and to predict the optical output power when the current drive and chip temperature change. The entire end-to-end free space optical (FSO) a link is disclosed, with a gain-switched temperature-dependent QCL as the transmitter. The FSO link device is made up of 90 injectors and active zones that produce light at 116μm. The gadget is powered by a variety of electrical inputs, including square, haversine, and tangential hyperbolic pulses. Gain switching generates brief pulses that travel 1500 m to reach the quantum well-infrared photodetector, which operates at the same wavelength as the source. For each input signal, the performance parameters signal-to-noise ratio (SNR), bit error rate (BER), and capacity are determined. Haversine input performs better under the minimum full width half maximum condition, with a BER of $7.8*10^{-5}$, a peak SNR of 14.56 dB, and a capacity of 4.89 bps/Hz at a cold finger temperature of 45K. Tangential hyperbolic input works well when peak power is used, with a minimum BER of $7.66*10^{-9}$, a peak SNR of 18.06 dB, and a capacity of 6.02 bps/Hz at a cold finger temperature of 45K.

13.12 Conclusion

QCLs are emerging as viable THz radiation sources. THz QCLs are rapidly gaining traction in a variety of applications such as standoff photoacoustic detection, atmospheric measurement, breath analysis for medical diagnostics, radio astronomy, high-sensitivity gas sensing and spectroscopy, security camera screening, and biomedical imaging, non-invasive techniques of industrial inspection of processes. Many materials, including clothes, polymers, biological materials, and packaging materials, are semi-transparent or transparent at THz frequencies, allowing for a wide range of security and public safety applications. The numerous high-quality references offered in this chapter will provide a thorough understanding of the various approaches for modeling QCL as well as the detailed numerical analysis that can be performed utilizing rate equations and reduced rate equations. With numerous papers concentrating on THz QCL-based Free Space Optical communications, the unlicensed spectrum offers a massive capacity for huge data transfers. The necessity for cryogenic cold fingers has been fully eliminated with the introduction of room temperature operated QCLs, and the device may now be used with ease.

References

[1] S. Houver et al., "Giant optical nonlinearity interferences in Terahertz quantum structures", 2020 45th International Conference on Infrared, Millimeter and Terahertz Waves (IRMMW-THz), 2021.

[2] Gary Agnew et al., "Frequency Tuning Range Control in Pulsed Terahertz Quantum-Cascade Lasers: Applications in Interferometry", IEEE JOURNAL OF QUANTUM ELECTRONICS, VOL. 54, NO. 2, APRIL 2018.

[3] H. W. Hubers et al., "Heterodyne Spectroscopy of Frequency Instabilities in Terahertz Quantum-Cascade Lasers Induced by Optical Feedback", IEEE Journal of Selected Topics in Quantum Electronics, 2017.

[4] K. Pierściński et al., "Optimization of cavity designs of tapered AlInAs/InGaAs/InP quantum cascade lasers emitting at 4.5μm", IEEE Journal of Selected Topics in Quantum Electronics, 2019.

[5] R.A. Khabibullin et al., "THz quantum cascade lasers based on GaAs/AlGaAs and HgCdTe material systems", 2020 International Conference Laser Optics (ICLO).

[6] Emilia Pruszyńska-Karbownik et al., "Optical Beam Characteristics of Quantum Cascade Laser with Mirrors Cleaned by Focused Ion Beam", 2018 International Conference Laser Optics (ICLO).

[7] R.A. Khabibullin et al., "The investigation of temperature degradation in THz quantum cascade lasers based on resonant-phonon design", 2018 International Conference Laser Optics (ICLO).

[8] P. Ashok et al., "Impedance characteristics of mid infra-red Quantum Cascade Lasers", Optics and Laser Technology, pp. 1–8, 2020.

[9] P. Ashok et al., "Performance evaluation of free space optical link by incorporating the device parameters of quantum cascade laser-based transmitter", Laser Physics Letters, Vol. 18, pp. 1–6, 2021.

[10] S. Gopinath et al., "Performance evaluation of free space optical link driven by gain switched temperature dependent quantum cascade lasers", Laser Physics Letters, Vol. 18, pp. 1–5, 2021.

Chapter 14

Design of Broad band Stacked Fractal Antenna with Defective Ground Structure for 5G Communications

G. Rajyalakshmi[1], and Y. Ravi Kumar[2]

[1]University College of Engineering, Osmania University, Hyderabad, Telengana, India
[2]Scintist-G (Retd) DLRL, DRDO, Hyderabad, Telengana, India
Email: [1]rajyalakshmi.gori@gmail.com, [2]dr.ravikumaryeda@gmail.com.

Abstract

A broadband fractal microstrip patch antenna has been designed for 5G communication applications. With dual layers and a defective ground structure, this antenna achieves a wider bandwidth and higher gain (DGS). In order to achieve broadband operation and high gain, the fractal elliptical patch is proposed to be printed on the FR4 substrate layer and coupled to the same FR4 substrate with DGS. The proposed antenna has an operating frequency of 3.5GHz and bandwidth of 2.7GHz to 5.3GHz. CSTMW 2018 is used for the analysis and simulation. Important parameters of the proposed antenna, such as S11, gain, and VSWR, have been simulated. The proposed structure has a compact design of 30 × 30mm^2. The proposed gain of the antenna is 7.56dB which is well suited for 5G applications. From the results, it is clearly shown that the antenna has excellent performance for 5G applications.

14.1 Introduction

In today's communication systems, small antennas are used. In addition, new multiband operating standards must be devised. Using a fractal approach to design convoluted geometries that result in smaller antennas is a reason to develop a solution for achieving compact, low profile antennas with multi-band and broadband characteristics while maintaining a low profile because of the two most common properties of fractal geometries. As a result, several research investigations have been conducted in order to build small multiband antennas that can serve a wide range of communication applications [1–2]. In [3,] a hexagonal fractal ring patch antenna for dual-band frequency coverage of 2.45 GHz Wi-Fi and 3.5 GHz WiMAX applications is presented. Miniaturization of FSS is another efficient way of boosting incident angle ranges. [5,] used dual-layered periodic arrays of fractal square loops to demonstrate broadband and wide-angle band stops FSS. [6] Exhibited an ultra-wideband closely coupled phased array antenna with integrated feed lines that had a low-profile. The aperture array is made up of planar element pairs with fractal geometry. These pairings are orthogonal to each other in each element for dual polarization. The design is made up of a series of densely capacitively coupled pairs of fractal octagonal rings. [7]shows that the sensor network designer who discovered a flat gain came up with a better new design. For the low profile cylindrical segment fractal DRA for wideband applications, [8,] presented several DRA feeding approaches and bandwidth augmentation techniques.[9] covers the development of novel wireless microstrip antennas based on modified Minkowski-like models. A tiny dual band microstrip BSF with fractal shaped resonators of different lengths is described in [10]. [11] Created, simulated, and implemented a new polygonal loop for passive UHF RFID applications. Increasing the number of sides on a fractal antenna enhances performance while reducing size, according to research. To increase the gain of a standard microstrip antenna, [12] created a Double-Octagon Fractal Microstrip Yagi Antenna (D-OFMYA). [13] Developed a high isolation UWB-MIMO antenna with a bandwidth of up to 8.6GHz based on a Minkowski fractal structure. Fractal geometry can assist small and multiband antennas and arrays, as well as high-directivity elements created without the usage of an antenna network. To achieve miniaturization, the antenna radiator was designed using a unique Sierpinski Knopp fractal geometry, and the antenna elements were positioned orthogonally to each other for isolation. [16] Includes an overview of fractal antenna designs with repeated patterns at different scales, as well as revisions to current state-of-the-art fractal antenna designs. A stacked two layer fractal microstrip patch antenna was invented by [17] to increase antenna characteristics. This research proposed a 5G communication dual-layer elliptical fractal antenna. We present a 2.7–5.2 GHz dual-layer fractal with a bandwidth of two layers. The suggested antenna has excellent gain and bandwidth due to the usage of a defective ground structure (DGS).

14.2 Introduction to the Fractal Concept

Fractals are named after the Greek word “Frangere,” which means “broken or uneven fragments”. It was created for the first time in 1975 by mathematician Benoit Mandelbrot. Because of its capabilities and competencies, fractal antennas are multiband, high gain, and low profile antennas that are employed in Wi-Fi packages. The practice of subdividing a shape into smaller copies of itself is known as iterative geometry. Traditional antennas operate on a single frequency band, necessitating a greater region for antenna coupling. Fractal geometry has been applied to a variety of fields, with positive results.

14.2.1 The fractal geometry

The self-similar structure of the fractal antenna, which is generated through an iterative process, distinguishes it from other types of antennas. In accordance with the simple mathematical shapes illustrated in the table below, fractal structures can be separated into two types: geometric and fractal.

Fractal structures with deterministic properties are known as deterministic fractal structures. These are made up of multiple miniaturized and rotated copies of the entire structure, such as the Sierpinski gasket, Von-Koch snowflakes, and Minkowski curve, among other things.

Fractal Structure with Random Elements: A fractal structure with random or irregular elements is a structure found in nature. The designs for these structures were chosen completely at random. Through the use of this method, it is possible to model and reproduce natural phenomena such as dendrites and lightning bolts, among others.

14.3 Antenna Design with a Single Layer Fractal

The single-layer elliptical fractal structure is discussed in this section. The used substrate is FR4, which has a dielectric constant of 4.3, a height of 1.6 mm, and a loss tangent of 0.02. Figure 14.1 depicts the geometry of the planned fractal’s zero, first, and second iterations. The zero-iteration structure is a single-layer elliptical structure, as shown in Figure 14.1(a), and Figure 14.1(b) depicts the first iterative fractal structure created by using a sub-elliptical structure as a starting point. Fig 14.1(c) depicts the second iterative fractal structure derived from the first iterative fractal structure using four basic fractal elliptical units. The high-level iterative fractal structure is finally designed. From the observations, all the iterative structures have the same operating frequency of 3.5GHz. The total size of the elliptical structure

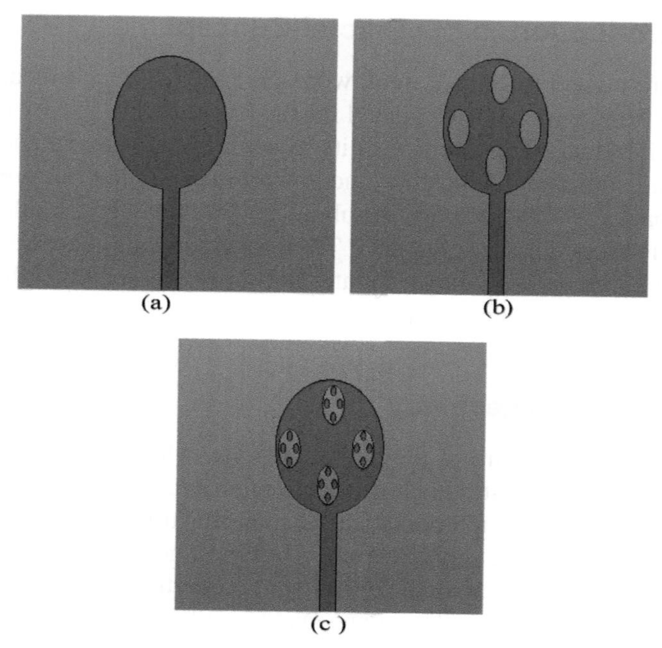

Figure 14.1 Design of the fractal geometries; (a) zero iteration, (b) First iteration, (c) second iteration.

is 30 × 30mm^2.The proposed antenna's parameters are Ws = 30mm and Ls = 30mm, Wf = 1.5mm and Lf = 11.2mm, and the proposed elliptical antenna's major and minor axes are a = 5mm, b = 7mm, c = 2mm, d = 4mm, and e = 0.25mm, f = 0.5mm, as shown in Figure 14.2. Figure 14.3(a) depicts the S11-Parameter of all iterations of the desired elliptical antenna. Figure 14.3(b) depicts the proposed antenna's surface current distribution, in which maximum current is distributed throughout the antenna. Figure 14.4 depicts the proposed antenna's gain details for each iteration, which improves between the first and second iterations.

14.4 Antenna Design with a Dual Layer Fractal

This section describes how to use proximity feed coupling to increase the bandwidth and gain of a dual-layer fractal antenna. The second substrate is the same as the first, with a dielectric constant of 4.3, a height of 1.6 mm, and a loss tangent of 0.02. The feed's width and length are 0.2mm and 13mm, respectively. Figure 14.5 depicts the dual-layer structure of the proposed element with etching (DGS) some portion from the ground, which provides a

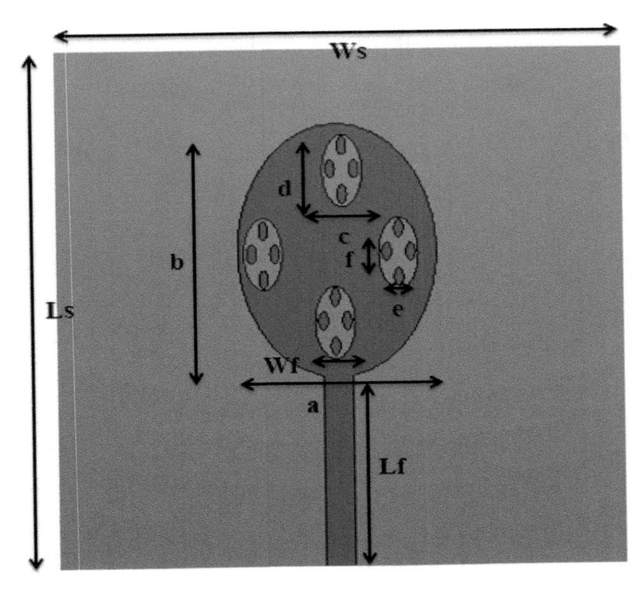

Figure 14.2 Dimensions of the single layer proposed antenna.

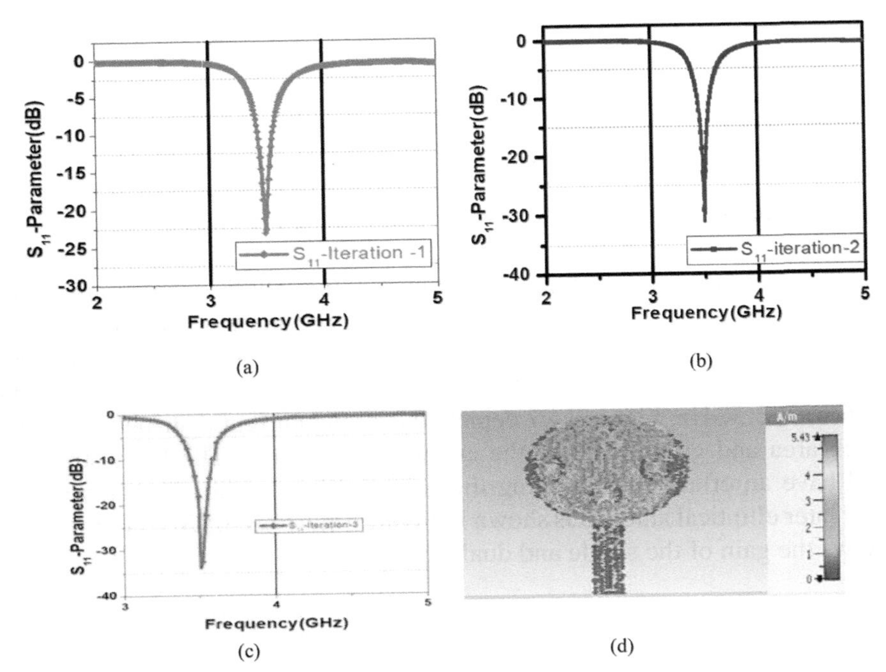

Figure 14.3 S-Parameter results a-c iterations (d) surface current distribution at 3.5GHz.

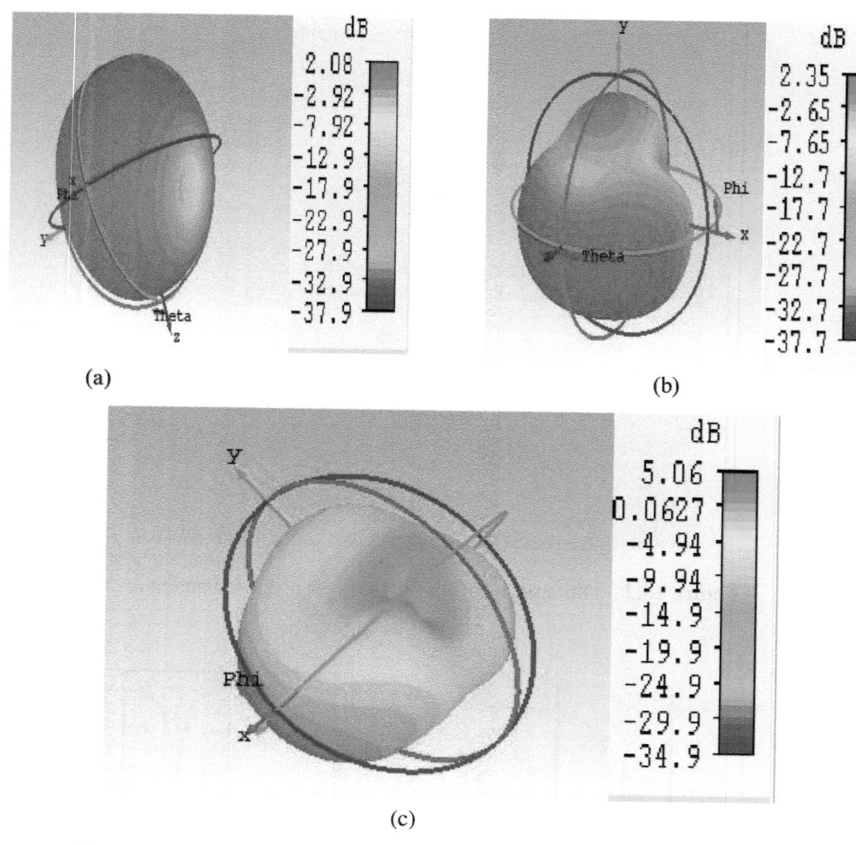

Figure 14.4 (a) Iteration-Zero gain (b) Iteration-1 gain (c) Iteration-2gain.

very low return loss -68dB and a very high gain 7.56dB when compared to the single-layer antenna. The dual-layer structure increases the bandwidth from 2.7GHz to 5.3GHz, which is nearly 3GHz. The simulated return loss and E-Plane and H-Plane with dual-layer are shown in Figure 14.6(a) and 14.6(b) at 3.5GHz. Figure 14.7 depicts the strong current distribution on the patch area and ground. When the antenna is activated. The DGS structure will have an effect on increasing the gain and bandwidth. The gain of the dual later elliptical antenna is shown in Figure 14.7(b) at 3.5GHz. Figure 14.8 shows the gain of the single and dual-layer antenna.

14.5 Conclusion

A novel study on the broadband and high gain dual layer patch antenna with DGS structure has been carried out in detail. The analysis was carried out

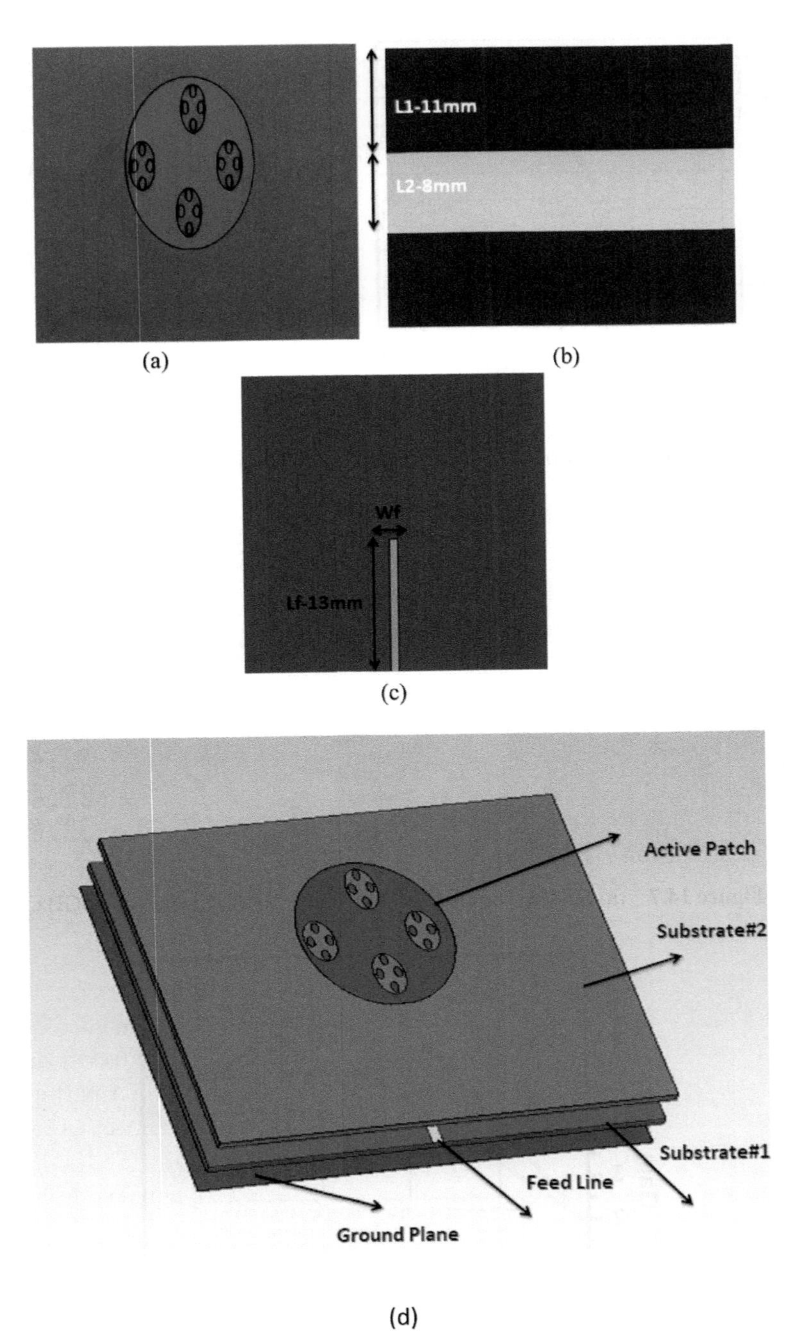

Figure 14.5 (a) Dual layer fractal front view (b) DGS back view(c) feed line (d) dual layer structure side view.

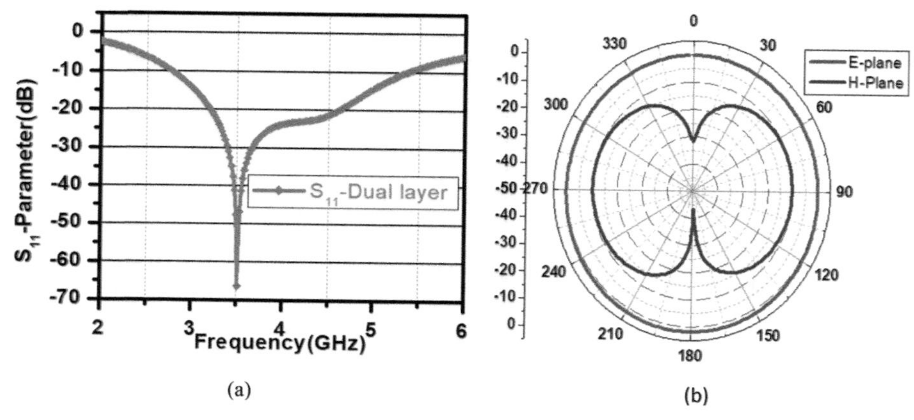

Figure 14.6 (a) S-Parameter result (b) radiation pattern at 3.5GHz.

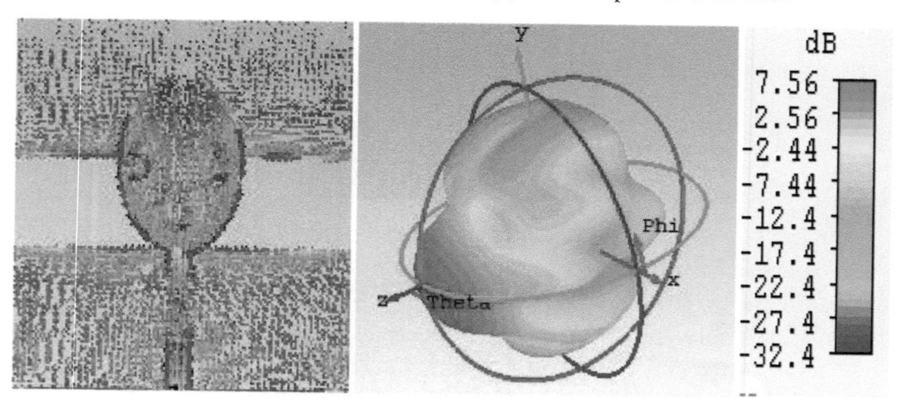

Figure 14.7 (a) Surface current distribution at 3.5GHz (b) Gain at 3.5GHz.

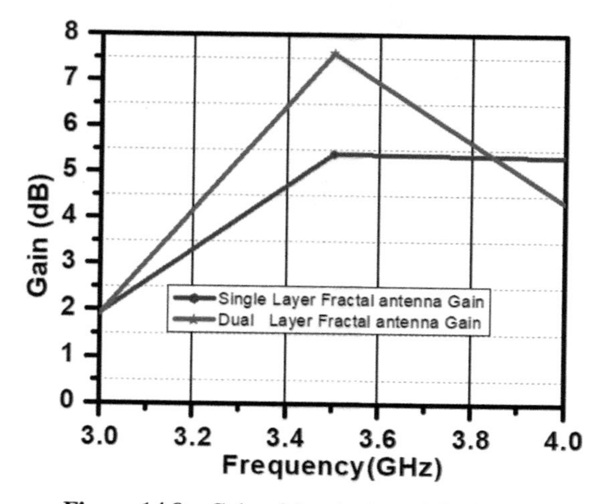

Figure 14.8 Gain of the single and dual layer.

Table 14.1 Compares the proposed work to the existing work.

Reference No	Frequency (GHz)	No of Layers	Return loss(–dB)	Gain	Bandwidth
12	5.8	single	–24.2	14.49	5.69–5.89
7	4.15 and 5.93	Single	–40, –13.5	>6	500MHz and 1GHz
17	10	Stacked dual layer	–34	–	7.5GHz – 14.4GHz
18	7.96, 9.22, 11.48	Double Fractal layer	–25	6	4.1GHz – 19.4GHz
Proposed work	3.5	Dual Layer Fractal	–68	7.56	2.7GHz – 5.3GHz

with a single and dual-layer fractal antenna. In that study the dual-layer fractal antenna return loss is –68dB and gain is 7.56dB. The designed fractal antennas were simulated using CSTMW 2018. In summary, the antenna may be recommended for use in the 5G operating frequency band.

14.6 Acknowledgement

We sincerely thank Mr.Sambasiva Rao, Scientist-E, RCI (DRDO), and Hyderabad for his guidance and support and also for providing resources to test the antennas with a network analyzer at an anechoic chamber. We also thank Prof Dr. D.Ramakrishna, Osmania University for CSTMW & HFSS software to generate simulations.

References

[1] H. K. Kan and R. B. Waterhouse, "Size reduction Technique for shorted patches", Electronics Letters, vol. 35, no. 12, 10th June 1999.

[2] Y.K Choukiker, S.K. Sharma, and S.K. Behera,"Hybrid Fractal Shape Planar Monopole Antenna Covering Multiband Wireless Communications With MIMO Implementation for Handheld Mobile Devices", IEEE Transactions On Antennas And Propagation, vol. 62, no. 3, March 2014.

[3] Kahina DJAFRI, Mouloud CHALLAL, Rabia AKSASA, Jordi ROMEU " Novel Miniaturized Dual-band Microstrip Antenna for WiFi/WiMAX Applications", published in 2017 IEEE.

[4] Ranga Y, Matekovits L, Weily A R, et al. "A low-profile dual-layer ultra-wideband frequency selective surface reflector." Microwave & Optical Technology Letters 55.6, 2013:1223–1227.

[5] Xianyou Xie, Wei Fang, Ping Chen * , Yin Poo and Ruixin Wu, "A Broadband and Wide-angle Band-stop Frequency Selective Surface

(FSS) Based on Dual-layered Fractal Square Loops,"published in 2018 IEEE.

[6] Eman O. Farhat, Kristian. Z. Adami† Yongwei Zhang ,Anthony K. Brown§ Charles. V. Sammut , "Ultra- wideband Tightly Coupled Fractal Octagonal Phased Array Antenna", 978-1-4673-5707-4/13/$31.00 ©2013 IEEE.

[7] Subhrakanta Behera ,Debaprasad Barad "A Novel Design of Micro strip Fractal Antenna for Wireless Sensor Network". 978-1-4673-6524-6/15/$ 31.00 © 2015 IEEE.

[8] Shubha Gupta, Poonam Kshirsagar, Biswajeet Mukherjee, "A Low-Profile Multilayer Cylindrical Segment Fractal Dielectric Resonator Antenna: Usage for wideband applications", IEEE Antennas and Propagation Magazine, 61(4), 55–63.Doi:10.1109/MAP.2019.2920424.

[9] Yaqeen S.Mezaal "New Microstrip Quasi Fractal Antennas: Design and Simulation Results", 978-1-5090- 1431-6/16/$31.00 ©2016 IEEE.

[10] Y. S. Mezaal, J. K. Ali, and H. T. Eyyuboglu, " Miniaturised microstrip bandpass filters based on Moore fractal geometry," International Journal of Electronics, pp. 1-14, 2014.

[11] Qusai H. Sultan and Ahmed M. A. Sabaawi, "Design and Implementation of Improved Fractal Loop Antennas for Passive UHF RFID Tags Based on Expanding the Enclosed Area", rogress In Electromagnetics Research C, Vol. 111, 135–145, 2021.

[12] Kamelia Quzwain,Alyani Ismail and Aduwati Sali, "A High Gain Double-Octagon Fractal Microstrip Yagi Antenna", Progress In Electromagnetics Research Letters, Vol. 72, 83–89, 2018.

[13] Yong Cai, Guangshang Cheng, Xingang Ren, Jie Wu, Hao Ren,Kaihong Song, Zhixiang Huang and Xianliang . "Highly Isolated Two-Element Ultra-Wideband MIMO Fractal Antenna with Multi Band- Notched Characteristics", Progress In Electromagnetics Research C, Vol. 116, 13{24, 2021.

[14] Anguera, J., A. And´ujar, J. Jayasinghe, V. V. S. S. S. Chakravarthy, P. S. R. Chowdary, J. L. Pijoan, T. Ali, and C. Cattani, "Fractal antennas: An historical perspective," Fractal and Fractional, Vol. 4, No. 1, 3, 2020.

[15] Rajkumar, S., K. T. Selvan, P. H. Rao, "Compact 4 element Sierpinski Knopp fractal UWB MIMO antenna with dual band notch," Microwave and Optical Technology Letters Vol. 60, No. 4, 10231030, 2018.

[16] Muthu Ramya C, R. Boopathi Rani, "A Compendious Review on Fractal Antenna Geometries in Wireless Communication". 978-1-7281-4685-0/20/$31.00 ©2020 IEEE.

[17] Jyotibhusan Padhi. Arati Behera, Muktikanta Dash, "Design of a Stacked Two Layer Circular Fractal Microstrip Antenna for X-band Application". 978-1-5090-3646-2/16/$31.00 ©2016 IEEE.

[18] Roman Kubacki *, Mirosław Czyzewski and Dariusz Laskowski, "Minkowski Island and Crossbar Fractal Microstrip Antennas for Broadband Applications". Appl. Sci. 2018, 8, 334; doi:10.3390/app8030334.

Chapter 15

Performance Analysis of T Shaped Structure for Satellite Communication

Jacob Abraham[1], R. S. Kannadhasan[2], R. Nagarajan[3], and Kanagaraj Venusamy[4]

[1]Associate Professor, Department of Electronics, B P C College, Piravom, Eranakulam, Kerala, India
[2]Assistant Professor, Department of Electronics and Communication Engineering, Cheran College of Engineering, Tamilnadu, India
[3]Professor, Department of Electrical and Electronics Engineering, Gnanamani College of Technology, Tamilnadu, India
[4]Control Systems Instructor, Department of Engineering, University of Technology and Applied Sciences-AI Mussanah, AI Muladdha,
Email: tjacobabra@gmail.com; kannadhasan.ece@gmail.com, https://orcid.org/0000-0001-6443-9993; krnaga71@yahoo.com, https://orcid.org/0000-0002-4990-5869;
Sultanate of Oman. kanagaraj@act.edu.om

Abstract

The proposed antenna is 30 × 30 mm in size. A 1.6 mm thick foam dielectric layer with a dielectric constant of 4.4 and a substrate thickness of 1.6mm is used as the dielectric layer. In the Ku band, simulating reflection coefficients allows for the computation of a broad bandwidth of 7.5GHz, 8.5GHz, and 9.5GHz, as well as a gain of 5.8dB, 6.5dB, and 7.5dB. Satisfaction elevation radiation may be used in satellite applications since it meets the balanced condition across a wide bandwidth. Antenna with a T-shaped construction.

15.1 Introduction

We occasionally demand a little suitable antenna to transport data to those smart home gadgets when we have a lot of smart devices for remote sensing and home automation in the current IoT age. The bulk of these devices is intended to make our lives easier. The authors show a variety of studies and explore how 2.4GHz Wireless Sensor Networks (WSNs) have become one of the most fascinating areas of research in recent years. Apart from that, there are a variety of interesting Microstrip antennas and array designs at 2.4GHz for Low Energy Bluetooth or RFID, but the focus of this research will be on a basic transmitting antenna design for the 2.4GHz radio, which is widely used for remote sensing and control. In the literature, there are many different 2.4GHz antenna designs to choose from. We wanted to contribute a bit to these regularly used antennas, thus we concentrated on 2.4GHz antennas. The goal was to improve the gain and bandwidth of a basic T-shaped inset fed patch antenna so that it may be used in multiple channel communications like WiFi or Low Energy Bluetooth, which is becoming more common in this era of IoT devices. [1]–[6]. In order to create a decent patch antenna design, it's crucial to collect a variety of simulation results. This research used Sonnet Suites, a planar electromagnetic modeling tool. Impedance and antenna bandwidth are the first factors to consider when creating a viable design. To retain input impedance, formulas on various antenna designs were used, and Network Analyzer was used to analyze this parameter. In recent years, mobile phones have reduced in size and gotten lighter in weight. Mobile phone components must be small, light, and have a low profile as a consequence of this trend. As a consequence, antennas used in mobile phones for personal communications must be small and essential, as the bandwidth and efficiency they provide may either enhance or limit system preference. Engineers must focus on the antenna system of the mobile phone in order to produce a gadget that performs better. The microstrip patch antenna is perfect for tiny and light phones since it is compact, light, and easy to produce [7]–[14].

15.2 T Shaped Structure Antenna

A microstrip patch antenna is a narrowband antenna made by etching the antenna element pattern in a metal trace bonded to an insulating dielectric substrate, with a continuous metal layer connected to the opposite side of the substrate as a ground plane. The low bandwidth of microstrip patch antennas is a significant disadvantage. There have been various successful attempts to enhance the bandwidth, gain, and efficiency of microstrip antennas. Some of these methods include changes in antenna size, dielectric constant, substrate thickness, and parasitic patches. The purpose of this research is to develop a microstrip patch antenna for LTE mobile terminals. This paper depicts the design of a T-shape patch microstrip. You can build an antenna by cutting four notches in a rectangular patch. By deleting slots from a patch, a microstrip antenna's gain, return loss, and bandwidth may be improved. The research provides an improved configuration of tiny T-shaped microstrip patch antenna design, modeling, fabrication, and testing on a 1.8mm FR-4 substrate. It also has approximately 18000 channels divided into different bands. A computer-simulated design model of the proposed T-shaped antenna is shown in Figure 15.1. The antenna consists of a portion of two standard rectangular patches that culminate in a T shape whose upper width is somewhat extended above the lower rectangular width, in contrast to earlier research with any complex antenna design.

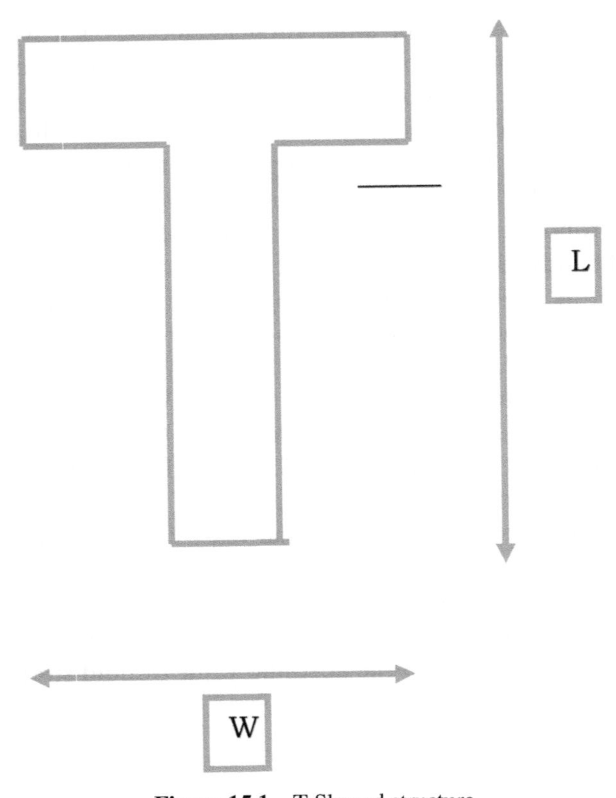

Figure 15.1 T Shaped structure.

To provide additional degrees of freedom for improving antenna performance in response to specific exposure uniformity criteria. Because of the natural suppression of edge currents that can be induced on the patch geometry, the antenna is fabricated on an epoxy FR-4 substrate with a dielectric constant of r = 4.4, resulting in lower fabrication costs and improved antenna performance, such as impendence bandwidth, gain, and better symmetry of the main beam, as well as lower side lobe levels. In general, the antenna profile is determined by the two adjustable lengths, width, and dimensions of the waveguide port. The patch antenna's length is 6.811 mm (L), its width is 9.603 mm (W), and the height of the substrate thickness is 1.6 mm, as illustrated in figure 15.1.

15.3 Results and Discussion

A T-shaped patch antenna was constructed and successfully replicated in the Ku band at 7.5, 8.5, and 9.5 GHz in this article. The simulated result is generated using computer simulation technology software. The proposed antenna satisfies the bulk of the criteria for satellite communication systems. In this work, a broadband microstrip rectangular patch antenna with a T-shaped antenna is built on an FR4 substrate with a thickness of 1.6 mm to achieve Ku band operating frequencies of 7.5, 8.5, and 9.5 GHz, which might be used for satellite communication and radar systems. The proposed antenna features a good elevation radiation pattern and a broad bandwidth impendence. Because the antenna is so simple and tiny, it might be used as an array in satellite communication systems for high-power operations elevation radiation. The recommended antenna has return losses of –22.65, –26.75, and –30.68dB, respectively, at 7.5, 8.5, and 9.5GHz, as shown in figure 15.2, with VSWRs of 1.5, 1.7, and 1.9, as shown in figure 15.3. Figure 15.4 shows the Gain, which is utilized to operate the different frequencies at 7.5, 8.5, and 9.5 GHz and is achieved at 5.8, 6.5, and 7.5dB. It is a measurement of an antenna's capacity to radiate in the desired direction with the least amount of antenna losses. For the whole solid angle coverage radiation, it is computed by multiplying the intensity of the radiation by the total input power to the antenna, as illustrated in figure 15.5. The two-dimensional radiation pattern may be seen using the variation of the absolute value of field strength or power

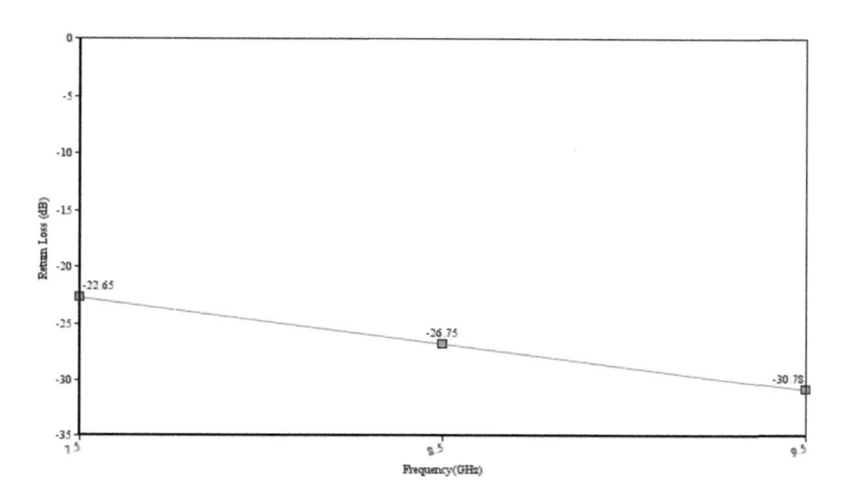

Figure 15.2 Return loss of the antenna.

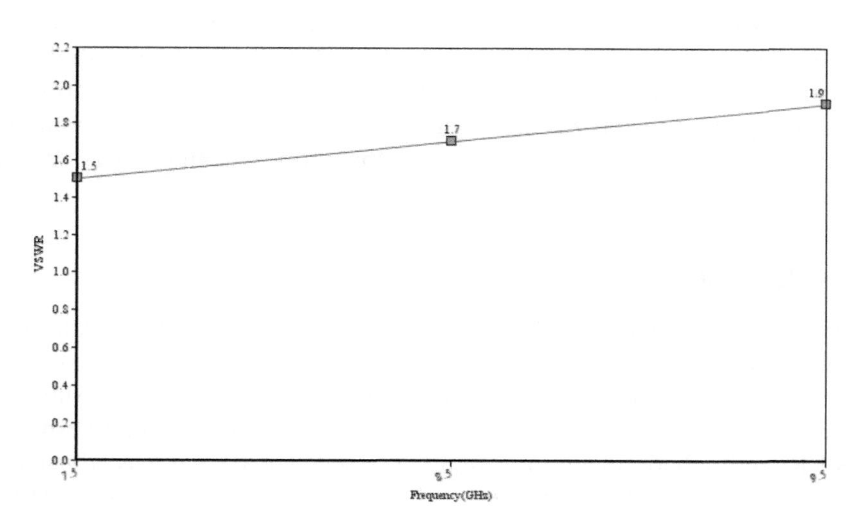

Figure 15.3 VSWR of the antenna.

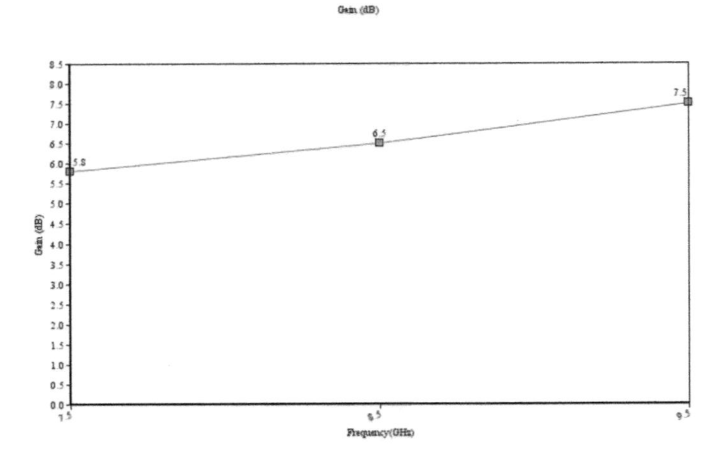

Figure 15.4 Gain of the antenna.

as a function of. The antenna's important Directional properties, such as HPBW, FNBW, Direction of Propagation, FBR, and so on, are shown on this curve. The proposed microstip antenna's directional qualities are affected by significant differences in the substrate thickness, dielectric constant, patch shape, effective length, Feed position, and W/L Ratio. The radiation pattern visualizes the antenna's propagation features for antenna optimization. The radiation pattern of the proposed antenna is seen in Figures 15.4 and 15.5.

15.4 Conclusion

HFSS (High Frequency Structure Simulator) software version 13.0, a Finite Element Method-based tool, was used to investigate the performance of a T-shaped triple-band microstrip patch antenna with FR4epoxy substrate operated by edge feed. The intended antenna's bandwidth, gain, return loss, VSWR, and radiation pattern were all assessed. The design was adjusted in order to get the best possible result. FR4epoxy was used as the substrate, which has a dielectric constant of 4.4 and is available in a range of thicknesses. The results show that a multiband antenna with a substrate thickness of 1.6mm may operate at frequencies of 7.5, 8.5, and 9.5GHz. Due to its frequency of operation and compact footprint, the recommended antenna is ideal for a wide range of devices used in WLAN, Wi-Fi, WiMAX, and other wireless and C-band applications.

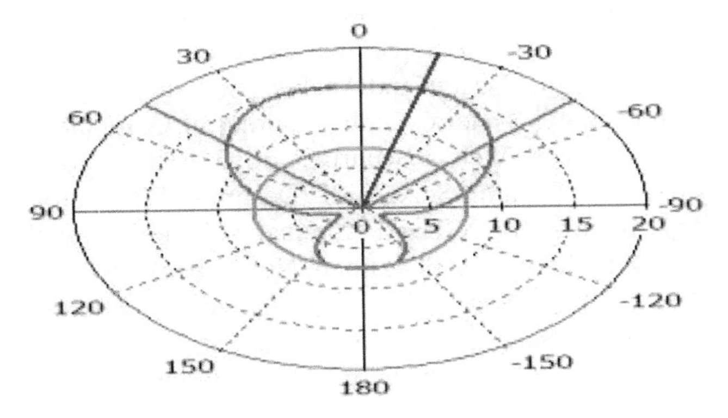

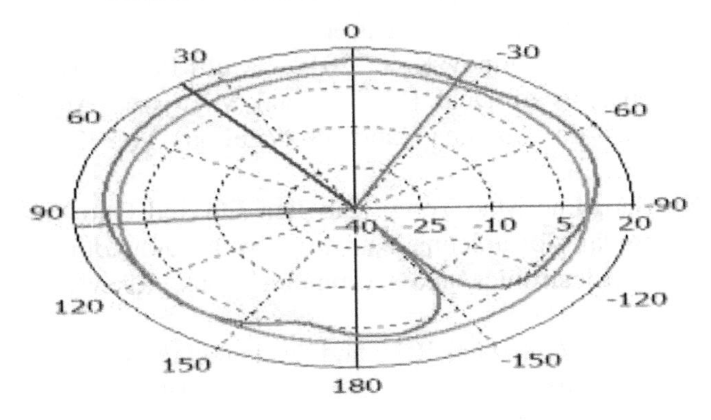

Figure 15.5 Radiation pattern of the antenna.

References

[1] A.Q. Khan, M. Riaz, A. Bilal, "Various types of antenna with respect to their applications: A review" International Journal of Multidisciplinary Science and Engineering, vol.7, no.3, March 2016.

[2] A. Mehta, "Microstrip Antenna" International Journal of Scientific & Technology Research, vol. 4, issue.3, March 2015.

[3] P. A. Ambresh, P. A. Hadalgi, P. V. Hunagund, "Effects of Slots on Microstrip Patch Antenna characteristics" International Conference on Computer and Electrical Technology, 978-1-4244-9394-1/11, March 2011.

[4] Y. L. Kuo, K. L. Wong, "Printed Dual T Monopole Antenna for 2.4/5.2 GHz Dual Band WLAN Application" IEEE Transaction on Antenna and propagation, vol.51, no. 9, September 2003.
[5] A. Kumar, S. Singh, "Design and Analysis of T shaped Microstrip Patch Antenna for the 4G System" Global Journal of Computer Science and Technology Network Web & Security, vol.13, issue 8, 2013.
[6] A. Goyal, M. R. Tripathi, S. A.Zaidi "Design & Simulation of Inverted T Shaped Antenna for X Band Application" International Journal of Computer & Technology, vol.4, no. 6, November 2014.
[7] KaushikS, DhillonS.S, 2013"Reactangular Microstrip Patch Antenna with U-Shaped DGS Structure for Wireless Applications" 5th International Conference on Computational intelligence and Communication networks, pp-27–31.
[8] Varadhan C., Pakkathillam J. Kizhekke, Kanagasabai M., Sivasamy R. Natarajan R. and Palaniswamy S. Kumar, 2013. "Triband Antenna structures for RFID Systems Deploying Fractal Geometry", IEEE Letters on Antennas and Wireless Propagation, Vol. 12, pp 437-440.
[9] [9] Behera S. and Vinoy K. J., 2012. "Multi-Port Network Approach for the Analysis of Dual Band Fractal Microstrip Antennas", IEEE Transactions on Antennas and Propagation, Vol. 60, No. 11, pp 5100–5106.
[10] Sun Xu-Bao, Cao Mao-Yong, Hao Jian-Jun and Guo Yin-Jing, 2012. "A rectangular slot antenna with improved bandwidth", International Journal of Electronics and Communications, Elsevier, Vol. 66, pp 465–466
[11] Chaimool S., Chokchai C., and Akkaraekthalin. 8th Nov 2012 "Multiband loaded Fractal Loop Monopole Antenna for USB Dongle Applications". Electronics Letters, vol. 48, No-23.
[12] S.Kannadhasan and R,Nagarajan, Performance Improvement of H-Shaped Antenna With Zener Diode for Textile Applications, The Journal of the Textile Institute, Taylor & Francis Group, DOI: 10.1080/00405000.2021.1944523
[13] G.C Nwalozie, C.C Okezie, S.U Ufoaroh, S.U Nnebe,"design and performance optimiation of a microstrip patch antenna ",International Journal of Engineering and Innovative Technology (IJEIT) Vol 2, Issue 7, pp. 380–387, January 2013
[14] S.Kannadhasan and R.Nagarajan, Development of an H-Shaped Antenna with FR4 for 1-10GHz Wireless Communications, Textile Research Journal, DOI: 10.1177/00405175211003167 journals.sagepub.com/home/trj, March 21, 2021, Volume 91, Issue 15-16, August 2021

Index

A
AOMDV algorithm 31

B
Bandwidth 185
BER 14
BPSK 22
BSS 2

C
Channel state information 114
Clustering algorithm 141
CMOS 60
communication system 31

D
Data loss 29
Decision Tree Classifier 149
DGS 185
digital signal processing 96
Digital VLSI 60
DSL broadband 17

E
ECG 2
Edge Detection algorithm 141
Electric Field Domains 179

F
Fast Independent Component Analysis 1
FBS 81
FDMA 85
Feasibility 32
FFT 99
FMU 81
FPGA 3
Fractal 188
Free Space Optical 181
Frequency Domain Subgroup Algorithm 111

G
GaAs/GaAlAs 177
Gain 185
Gaussian distributions 2
GPS systems 48
GSM standard 14

H
Hardware Compatibility List 51
HFSS 202

I
ICA 3
IFFT 100
ISI 15

K
K-Nearest Neighbour Classifier 149

L
LAMDA 116
Laser Feedback Interferometry 176
LDR 143
Light Dependent Resistor 140
Long-Term Evolution (LTE) 15
Loss of packets 37

M
MATLAB 9
Maximum likelihood of Frequency 88
MIME 56
MIMO-OFDM-BICM 20
MIMO 59
MMSE algorithm 3
MRI coils 123
MSE 56
MUD 29
Multiple beam parasitic array radiator 164

N
Naïve Bayes Classifier 150

O
OFDM method 15
OPEX 72
Optical fiber 18

P
Packet Transmission 29
Parasitic 169
PFDMA 71
Photovoltaic 140
Principal Component Analysis 2

Q
QAM 16
QCL 176
QoS 47
QPSK 22

R
Radiation Pattern 203
Radio Frequency 124
Radio resource management 113
Retransmission 41

S
SAR 9
SC-FDMA 74
Schnorr-Euchner 59
SDR 9, 56
SE algorithms 59
SE decoding 62
Servomotor 142
SINR 48
SIPO 99
SIR 9
SOC architecture 64
SOCP problem 49
Solar Panel 144
Supervised learning 148
Support Vector Machine 150

T
TDMA 112
Terahertz quantum cascade-lasers 175

U
UWB 186

V
VB decoding 61
VBLAST 83
VHDL 4
Viterbo-Boutros 59
VSWR 165

W
Wi-Fi CDMA 13
WiMAX 164
wireless PAN network 34
Wireless signals 49
WLAN 163

About the Editors

S. Kannadhasan is working as an Assistant Professor in the Department of Electronics and Communication Engineering at Cheran College of Engineering, Karur, Tamilnadu, India. He has completed his research in the field of Smart Antenna for Anna University. He is ten years of teaching and research experience. He obtained his B.E in ECE from Sethu Institute of Technology, Kariapatti in 2009 and M.E in Communication Systems from Velammal College of Engineering and Technology, Madurai in 2013. He obtained his M.B.A in Human Resources Management from Tamilnadu Open University, Chennai. He obtained his PGVLSI in Post Graduate Diploma in VLSI design from Annamalai University, Chidambaram in 2011 and PGDCA in Post Graduate Diploma in Computer Applications from Tamil University in 2014. He obtained his PGDRD in Post Graduate Diploma in Rural Development from Indira Gandhi National Open University in 2016. He has published around 18 papers in the reputed indexed international journals and more than 125 papers presented/published in national, and international journals and conferences. Besides he has contributed a book chapter also. He also serves as a board member, reviewer, speaker, session chair, and advisory and technical committee of various colleges and conferences. He is also to attend various workshop, seminars, conferences, faculty development programmes, STTP, and Online courses. His areas of interest are Smart Antennas, Digital Signal Processing, Wireless Communication, Wireless Networks, Embedded Systems, Network Security, Optical Communication, Microwave Antennas, Electromagnetic Compatability, and Interference, Wireless Sensor Networks, Digital Image Processing, Satellite Communication, Cognitive Radio Design, and Soft Computing techniques. He is a Member of IEEE, ISTE, IEI, IETE, CSI, IAENG, SEEE, IEAE, INSC, IARDO, ISRPM, IACSIT, ICSES, SPG, SDIWC, IJSPR, and EAI Community.

R. Nagarajan received his B.E. in Electrical and Electronics Engineering from Madurai Kamarajar University, Madurai, India, in 1997. He received his M.E. in Power Electronics and Drives from Anna University, Chennai, India, in 2008. He received his Ph.D. in Electrical Engineering from Anna

University, Chennai, India, in 2014. He has worked in the industry as an Electrical Engineer. He is currently working as a Professor of Electrical and Electronics Engineering at Gnanamani College of Technology, Namakkal, Tamilnadu, India. He has published more than 70 papers in International Journals and Conferences. His research interest includes Power Electronics, Power System, Communication Engineering, Network Security, Soft Computing Techniques, Cloud Computing, Big Data Analysis, and Renewable Energy Sources.

Alagar Karthick working as an Associate professor in the Electrical and Electronics Engineering department at KPR Institute of Engineering and Technology, Coimbatore, Tamilnadu, India. He has published more than 30 International Journals and is also a reviewer for various journals such as Solar Energy, Fuel, Journal of Cleaner Production, Heliyon, Building services Engineering research and Technology. He received his Doctor of Philosophy in the field of Building Integrated Photovoltaic (BIPV) from Anna University, Chennai in 2018. He received his Master's degree in Energy Engineering and a bachelor degree in Electrical and Electronics Engineering. He has received Best Paper Award for his research articles on Biomass conversion. His research area includes Solar Photovoltaic, Bioenergy, zero energy buildings, Energy with Artificial Intelligence, Machine learning, and Deep learning algorithms.

Aritra Ghosh is a Lecturer at the Renewable Energy, University of Exeter, Penryn, UK. Prior to joining this position, he was Post-Doctoral Research Fellow in the same department. Previously, he worked as a Research Associate at The Centre for Industrial and Engineering Optics, Technological University, Dublin, Ireland. He has a Ph.D. in Building Engineering from Dublin Energy Lab, Technological University, Dublin, Ireland. He is a member of the Renewable Energy Group in the College of Engineering, Mathematics and Physical Sciences and also part of the interdisciplinary Environment and Sustainability Institute.

9788770227766